CROSSING THE RIVER

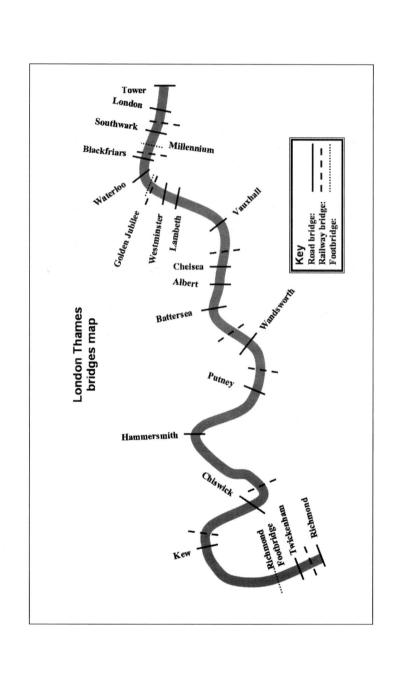

London Thames bridges map

Key
Road bridge:
Railway bridge:
Footbridge:

Tower
London
Southwark
Millennium
Blackfriars
Waterloo
Golden Jubilee
Westminster
Lambeth
Vauxhall
Chelsea
Albert
Battersea
Wandsworth
Putney
Hammersmith
Chiswick
Richmond Footbridge
Twickenham
Richmond
Kew

CROSSING *the* RIVER

The History of London's Thames River Bridges
from Richmond to the Tower

BRIAN COOKSON

MAINSTREAM
PUBLISHING
EDINBURGH AND LONDON

Dedicated to
Richard and Sarah,
and their families

First published in Great Britain in 2006 by
MAINSTREAM PUBLISHING COMPANY
(EDINBURGH) LTD
7 Albany Street
Edinburgh EH1 3UG

ISBN 1 84018 976 2

A catalogue record for this book is available from
the British Library

Typeset in New Baskerville and Frutiger

Printed in Great Britain by
William Clowes Ltd, Beccles, Suffolk

Acknowledgements

I wish to thank the staff of the British Library and of all the local libraries which have been so helpful to me in my research. I would especially like to mention Jeremy Smith of the Guildhall Library, City of London; Jane Baxter of the London Borough of Richmond upon Thames Local Studies Collection; Jane Kimber of the Hammersmith and Fulham Archives and Local History Centre; and Meredith Davis of the Wandsworth Local History Service, who have given generous help in providing the fascinating illustrations of historical bridges. The other local libraries whose help I wish to acknowledge are the City of Westminster Archives Centre, the Kensington and Chelsea Local Studies Collection, the Chiswick Library Local Studies Collection and the Lambeth Archives. Without the dedication of so many people who have collected and catalogued the magnificent range of local history material held in these libraries, this book would not have been possible. I would also like to express my thanks to the following for providing specific expert information: Patrick Connolly of the Marine Support Unit of the Metropolitan Police; John Ormsby, Chairman of the Strand on the Green Association; and Dana Skelley, Head of Central London Street Management. On a personal basis, I wish to thank Carol Smith and my agent, Charlie Viney, for putting me up to writing this book. Finally, I am proud to acknowledge the help of my daughter, Sarah, who has spent many hours improving the quality of the text, and of my wife, Susan, who has spent almost as long proofreading it.

Contents

Preface

Thousands of us cross over the River Thames in London every day. Most often we are too concerned with traffic congestion and the need to reach our destination to notice the structures of the bridges we use, let alone consider their history or the efforts and dangers that went into building them. However, as a London Blue Badge Guide, I am frequently made aware that when people do have time to look around, it is the views of the river from the bridges and many of the historic bridges themselves that most inspire them. London's Thames bridges constitute an irresistible subject of study, combining history with the romance of the imposing structures bestriding the powerful flow of England's longest river. There is something about the concept and appearance of a bridge that excites the human imagination. John Betjeman once wrote that it is difficult to make a bridge look ugly, although he did add wryly that this was achieved in the case of the iron railway bridge over the Thames at Charing Cross.

Lengthy books have been written about individual Thames bridges. Other books cover all the bridges from the source to the sea, or include other types of crossing, such as tunnels and ferries, and so have to limit their descriptions of each bridge to the basic facts. In this book, I have devoted a full chapter to each of the main bridges or groups of bridges on the tidal Thames within the area of Greater London, starting at Richmond and ending at Tower Bridge. Each chapter covers the historical background, why the bridge was built, problems in obtaining finance and approval, bridge design and construction, and, finally, subsequent developments up to the present day. I have also included historical illustrations as well as current photographs to show how the bridges and the river itself have changed over the years.

The book is aimed at the enquiring layperson rather than the professional engineer. I do, however, include what I hope will be intelligible descriptions of some of the technical aspects of bridge design along with the important statistics on each bridge in the appendices.

The diversity and vibrancy of modern London seems to be reflected in the idiosyncratic variety of its river bridges, from the eighteenth-century classicism of Richmond Bridge and the nineteenth-century Gothic extravaganza of Tower Bridge to the streamlined elegance of the twenty-first-century Millennium Bridge. My aim is to increase the reader's knowledge and appreciation of these bridges, their rich history and the people who built them, and thereby enhance the pleasure of experiencing the bridges, whether at leisure or stuck in a traffic jam.

Illustrations

Plates

All the illustrations are from the collections of Susan and Brian Cookson, except for those listed below, for which acknowledgement for the permission to reproduce them is gratefully made:

Sylvia Claydon (p. 36)

Guildhall Library, City of London (pp. 165, 175, 186, 257, 263, 273, 275, 283, 289)

Guildhall Art Gallery, City of London (p. 229)

London Borough of Hammersmith and Fulham (pp. 81, 85, 99, 104)

London Borough of Lambeth Archives Department, www.lambethlandmark.com (pp. 146, 156, 196, 208, 210)

London Borough of Richmond (pp. 31, 53, 56)

Wandsworth Local History Service (pp. 119, 133)

Introduction

Half a million years ago during one of the many ice ages that beset the British Isles, glaciers dammed the flow of a massive river to the north of present-day London. This caused the river to burst through the Goring Gap in the Chiltern Hills, pursuing its way to the North Sea roughly along the course of the current River Thames. Archaeological excavations provide evidence of many prehistoric human settlements on the north and south sides of the river. Even more fascinating are finds of elephant and hippopotamus bones in Trafalgar Square. These animals would have been able to cross the river by wading or swimming, and people would doubtless have used fords and boats to do the same from early times.

Until recently, it was assumed that the ancient Britons had not built any bridge crossings and that the first bridge to be constructed over the Thames was the Roman wooden bridge sited just downstream of today's London Bridge. However, in 2001 Channel 4's *Time Team* in conjunction with the Museum of London Archaeology Service (MoLAS) investigated some wooden stakes that had emerged from the river-bed upstream of Vauxhall Bridge and established that they were almost certainly the remains of a Bronze Age walkway over the Thames. This early bridge must have disappeared by the time Julius Caesar invaded Britain in 55 BC, as Caesar describes in his *Commentaries* how his army had to ford the Thames – the name given to the river by Caesar himself. He states that it was fordable in only one place, and although historians believe this was at Westminster or Brentford, there is no way we can be sure exactly where it was. What is certain is that the Thames was much wider, and in places shallower, than it is today and that much of the flood plain through which it flowed was marshy. The Romans finally conquered Britain in AD 43 and established Londinium on the firm stretch of high ground on the north bank of the river, opposite a low area of dry ground on the south bank. There they built a wooden bridge across the river just downstream of the present London Bridge.

The fate of the bridge after the Romans left Britain in 410 is unknown. From the tenth century, records start to appear with references to a wooden London Bridge at this site. In 1209, the wooden bridge was finally replaced by a stone one – the world-famous inhabited Old London Bridge with its houses, shops and even a chapel. Supported on 19 arches, amazingly it lasted over 600 years. It has often been said that the history of London is inextricably bound up with the river and, as we will see, it is also strongly reflected in Old London Bridge and its successors. This was the only bridge over the Thames in the central London area until the completion of Westminster Bridge in 1750. A wooden bridge had in fact been built in 1729 at Putney, but this was not really part of London at the time. The overcrowding of Old London Bridge suggested there was a dire need for a new bridge from at least the seventeenth century, when London started its massive expansion to the west.

The reason why it took so long to construct any other bridge was down to politics and finance rather than need or technology. Although Old London Bridge was always the preferred method of crossing between the north and south banks of the Thames, it also presented difficulties for those who used the river for transporting people or cargo. For centuries, watermen had offered their services from the many stairs on both sides of the river, taking people across the Thames as well as from place to

16

place along its banks. In general, river transport was popular well into the nineteenth century compared with the dirty, smelly and often dangerous roads. However, Old London Bridge proved especially dangerous to river traffic because of the swift flow of the tide through the narrow gaps between its arches, and many watermen and their passengers drowned in the passage. So the watermen saw bridges as obstructions, and to protect their interests they formed the powerful Company of Watermen, which received the Royal Charter in 1555. In addition, the City Corporation, which owned Old London Bridge, wished to preserve its monopoly on a bridge crossing so as to prevent trade from moving westwards. A combination of these vested interests managed to delay the approval of a new bridge at Westminster for nearly a century.

Once the precedent of a new crossing had been established at Westminster in 1750, several stone bridges across the Thames were approved and built in the eighteenth and early nineteenth centuries, including Robert Mylne's Blackfriars and James Paine's Richmond bridges. A great change occurred in bridge building with the coming of the Industrial Revolution and the construction of the world's first iron bridge, over the River Severn at Coalbrookdale, in 1779. Following the construction of London's first cast-iron bridge at Vauxhall in 1816, there was a profusion of bridge building using a variety of materials including cast iron, wrought iron, traditional stone and, later in the nineteenth century, steel. The coming of the railways contributed to this activity although it must be said that most of the railway bridges were utilitarian rather than beautiful. Many of London's bridges were initiated and financed by private enterprise and tolls were charged to recoup the investment. In 1879, the last bridges to charge tolls were finally bought out and all came under some form of public ownership.

The result of all this bridge building was that by 1900 there were eighteen road bridges, nine railway bridges and two footbridges over the Thames in London. As with London itself, there was no strategic plan and the pattern of crossings was somewhat haphazard. Moreover, all the road bridges had been built for the age of the horse and so most were inadequate for modern traffic. The other major problem was that, despite the involvement of famous engineers such as Brunel, Rennie and Bazalgette, most of the bridges failed to stand up to the fierce battering they received from the ebb and flow of the twice-daily tides, which reach speeds of up to 14 mph. Up until 1832, the tidal effect would have been much less severe because Old London Bridge, with its 19 arches, acted as a sort of weir. In fact, the flow of the tide was reduced to such an extent around Old London Bridge that the Thames often froze over in winter, leading to a succession of Frost Fairs, during which booths were set up on the frozen river, oxen were roasted and printing presses produced certificates for customers to record their attendance. The year 1814 saw the last of the Frost Fairs. Rennie's five-span replacement London Bridge of 1831 no longer slowed the tide enough for the river to freeze over, nor indeed to protect the upstream bridges from the tide's onslaught.

Apart from the eighteenth-century Richmond Bridge, made of stone, and the much later, nineteenth-century Albert and Tower bridges, all these early bridges have had to be rebuilt over the years, sometimes more than once. Often there has been considerable public protest when these historic structures have had to be pulled down. For instance, in the 1940s, there was outrage when the decision was taken to rebuild Rennie's popular Waterloo Bridge, which had been described by the sculptor Canova as the noblest bridge in the world. However, like the others, it finally had to come down and be replaced by today's more stable and traffic-friendly bridge of reinforced concrete.

Tower Bridge was completed in 1894. After that, apart from the Millennium Bridge of 2001, no new bridges have been built over the Thames in the central London area, and only a few have been built further upstream. When looking back at what happened to the numerous Thames crossings in Greater London, we are more amazed by the longevity of Old London Bridge – the very first stone bridge – than we are by the much shorter lives of its followers. Bridge construction presents many difficult problems to the engineer, especially when crossing a fast-flowing river like the Thames. Even in modern times, mistakes are made: for instance, the famous wobble experienced on the Millennium Bridge by the crowds who walked across at its opening in 2001. Thus it is particularly impressive that Old London Bridge lasted over 600 years despite the much more primitive technology available in medieval times.

Many of the engineering considerations involved in bridge building are technical and of little interest to the layperson. However, it is worth mentioning some major decisions that had to be made in designing these London bridges. Choice of material has been important, but this has depended on the technical advances of the day. Roughly speaking, it is true that stone was used until the end of the eighteenth century, iron and, later, steel in the nineteenth century, and in the twentieth century there was increasing use of reinforced and pre-stressed concrete. In general, improved materials allowed longer spans to be designed. The longest spans are usually found in suspension bridges, the first of which in London was William Tierney Clark's Hammersmith Bridge, completed in 1827, with a central span of 422 feet. One clear advantage of a longer span is that it enables river traffic to pass through more freely, but unfortunately, in the case of Hammersmith Bridge the road deck was slung so low that tall ships could only pass through at low tide. The benefit of a long span has to be weighed against the cost, as it is usually expensive to build. Design of the spans has to be considered along with the choice of the type of bridge (arch, beam, suspension, cantilever, cable stay, bascule and so on). All of these types are represented on the Thames. Brief descriptions of each type are contained in the appendix on Bridge Basics.

Perhaps the most crucial engineering decision was how to build the foundations of the bridge supports. Although the Thames is quite shallow at low tide, the difference in depth between low and high tide is about 21 feet at London Bridge and, as mentioned above, the ebb and flow of the tide can reach the considerable speed of 14 mph. The river-bed is mainly clay with variable coverings of gravel and sand. Clay is about the worst type of ground on which to build any foundations. It is estimated that it can bear about 40 times less weight than rock. Therefore it is essential to drive down deep into the river-bed to provide as firm foundations as possible for the bridge piers. This has been done by driving wooden piles into the river-bed for supports, as with Old London Bridge. For the longer spans of the later bridges, this method would not have provided a firm enough base. The two main techniques used were cofferdams and caissons. Cofferdams are made by driving wooden piles or, more recently, sheet steel, into the river-bed to form an enclosed space which can be pumped dry and filled with concrete or other strong material to provide the foundations on which to build the piers. A caisson is a sort of prefabricated cofferdam which has to be sunk into the river-bed either by dredging or excavation.

Once the bridge is completed it is subject to the problem of scour, whereby the flow of the water carries away material from the bridge supports over the years. As we will see, inadequate foundations have been the cause of the demise of many of London's Thames bridges. With the greater understanding of these issues available to the builders of the replacement bridges, it is to be hoped that the current structures will prove more long-lasting.

Today, there are 18 road bridges, 9 rail bridges and 3 footbridges over the Thames between Richmond and the Tower of London (mainly in similar positions to the bridge landscape of 1900). In addition, there are 15 tunnels carrying foot passengers, road vehicle traffic or London Underground trains under the Thames in the Greater London area, although many of these are to the east of Tower Bridge. No other major city has so many river crossings. It seems that land traffic interests have now won a complete victory over the proponents of river traffic. Although this trend has gathered momentum since the eighteenth century, many people regret it has gone so far and that the Thames is so little used as a highway today.

Looking to the future, it is likely that most bridge-building activity will be to the east of Tower Bridge, as there are plans to expand housing and stimulate economic growth in these underdeveloped areas. Strategic proposals are being considered, including one for a crossing at Thamesmeade, but so far no decision has been made. The following chapters examine London's river crossings, starting at the oldest existing bridge at Richmond in the west and working eastwards, ending at the most dramatic of all – Tower Bridge.

CHAPTER 1

Richmond and Twickenham

> Of all the stately works of man that we can enjoy as we
> voyage up the river to Oxford, there are three that stand
> out from all the others. These are Windsor Castle,
> Hampton Court and Richmond Bridge. Built of white
> stone of five arches which increase in height and span to
> the centre arch and crowned with stone balustrades and
> supported by rounded buttresses, this bridge of 1780 [*sic*]
> is indeed a thing of beauty.

This quotation from an undated article by Mr Donald Maxwell[1]
encapsulates the sense of aesthetic pleasure experienced by all
who see this elegant Palladian structure which spans the Thames
in the beautiful setting of Richmond riverside. Although only
the seventh bridge to be built on the lower reaches of the
Thames, it is the oldest remaining structure, as all the other
earlier ones have had to be replaced. Over the years, Richmond
Bridge has proved inadequate to convey all the traffic requiring
to cross the river in the area and has been supplemented by
three further bridges, which are designed to be functional
rather than beautiful. The first of these is the Richmond Railway
Bridge, built in 1848. This was followed by the Richmond
Footbridge, Lock and Weir in 1894 and the Twickenham Road

Bridge in 1933. Each proved controversial at the time, but today they are accepted as essential to the local economy and the preservation of the environment.

Richmond Bridge

Richmond Bridge replaced a ferry which from medieval times had provided a crossing for horse-drawn vehicles and pedestrians at about the same location on the river. The first mention of the ferry dates from 1443, in the reign of Henry VI, but it was almost certainly in existence from the time of Edward III's development of the Manor of Shene where he built himself a palace in the previous century. The ferry was always the property of the Crown and was leased to servants of the Crown or royal favourites to run it. Usage of the ferry will have increased considerably after Henry VII rebuilt the old palace, which was severely damaged by fire in 1499, and at the same time changed the name of the manor from Shene to Richmond, after his estates in Yorkshire. We even have fascinating records of his son Henry VIII's expenses, which indicate that he regularly spent money on the ferry. One record for December 1537 reads: 'Paid to Perkins of Richmond for the ferrying of the Princess and her servants arriving from Windsor – six shillings.' As the ferry, though leased, belonged to the monarch, this seems an uncharacteristically just act by the old autocrat. The princess referred to was the future Elizabeth I when she was four years old. She herself will doubtless have used the ferry often, as Richmond was her favourite palace and it was there that she died in 1603. There is a sad parallel between the deaths at Richmond of Elizabeth I and Edward III. Both had their rings cut off their fingers – Edward's by his thieving mistress, Alice Perrers, and Elizabeth's so that it could be delivered into the hands of James VI of Scotland to ensure the succession.

During the seventeenth and eighteenth centuries, Richmond developed into a thriving and fashionable town, although Henry VII's magnificent palace became neglected and was pulled down. The area kept its royal connections, however, and was the favourite country resort of George II and Queen Caroline. The

Queen built the imposing terrace on Richmond Green known as Maids of Honour Row for her ladies-in-waiting. Sir Joshua Reynolds, first president and founder of the Royal Academy, and Johann Christian Bach, the composer, both lived here for some time. Richmond also became a favourite riverside destination for tourists from London as it could be reached by coach in less than three hours. Chalybeate springs were found on Richmond Hill and a spa was developed there. Unfortunately, it soon became too popular, especially with rowdy groups who did not confine themselves to imbibing the health-giving waters. Mrs Susanna Houblon, the daughter of the first governor of the Bank of England, lived nearby on Richmond Hill. She bought the main buildings in 1763 and closed down the spa, which had become a nuisance to the local inhabitants. In 1768, the Theatre Royal was built on Richmond Green, replacing an older theatre on Richmond Hill. It was opened by David Garrick and attracted some of the most famous actors of the day, including Edmund Keane and Sarah Siddons.

Whereas Richmond was in the county of Surrey, Twickenham on the opposite side of the river was in the county of Middlesex. The Middlesex bank was less developed, but much favoured by aristocrats, artists and writers. Alexander Pope was among the first to build himself a villa here, in 1719. Later, Henrietta Howard, the mistress of George II when he was Prince of Wales, built a Palladian villa at Marble Hill and Horace Walpole, author and son of Prime Minister Robert Walpole, designed the extraordinary Gothic villa at Strawberry Hill. Pope's villa has long since vanished apart from the rather sad remnants of his famous grotto, but Marble Hill and Strawberry Hill still survive. Of the several artists who lived in Twickenham at this time, two were very much connected with the Thames and its bridges – Samuel Scott and his pupil William Marlow, who both painted central London river scenes in the style of Canaletto.

As a result of the developments here on both banks of the Thames, the need for a bridge to replace the ferry was becoming overwhelming. Horace Walpole records a number of occasions on which he had problems crossing the river by the ferry. Once, after dining in Richmond, he was forced to travel to Kew to cross back to Strawberry Hill via the new wooden bridge there because the river was too swollen for the Richmond ferry to operate. On another occasion, he did use the ferry but the darkness of the night, the rapidity of the current and the drunkenness of the bargemen nearly resulted in disaster. The first person to take action was none other than William Windham, to whom George II had granted the lease on the ferry until 1798. Windham had been the sub-tutor to the King's younger son, the Duke of Cumberland, who became notorious as the 'Butcher of Culloden', as he massacred the defeated Jacobites at the end of the 1745 rebellion. Windham was also husband of one of the King's former mistresses and therefore doubly qualified as a servant of the Crown. He had sub-let the ferry to Henry Holland and saw an opportunity to make a profit for himself by building a bridge to meet the increased demand for crossing the river. In 1772, he proposed a parliamentary Bill to allow the construction along the course of the ferry of a wooden bridge with nine arches, the design of which is still held in the British Museum.

The proposal caused uproar among the local inhabitants, which was typical of the many campaigns which have disturbed the apparent calm of Richmond's riverside environment over the years. A group was set up to fight the proposal for a variety of reasons. The inhabitants were incensed that the profits would accrue to a single individual, William Windham, and exclude other potential investors; and they wanted the bridge to cross the river at Water Lane near the centre of the town, where the approach was much less steep than at Ferry Hill as proposed by Windham. However, their main fury was directed at the design of the bridge and its construction in wood. In a letter to the *Lloyd's Evening Post* of 18 February 1772, an anonymous writer railed, 'What a cat-stick building must this be . . . Methinks I heard Old Thames groan to be so vilely strode.' William Windham seems to have buckled under this pressure, as he withdrew his Bill and left the field open for the inhabitants to put forward their alternative proposal, which formed the basis of the Act of Parliament which received Royal Assent on 1 July 1773.

The Act nominated 90 commissioners who were to be responsible for building and maintaining a bridge of stone construction. The commissioners included the landscape gardener Lancelot 'Capability' Brown, the writer Horace Walpole, the actor David Garrick and Sir Charles Asgill, the local MP and a former Lord Mayor of London, who had recently presided over the removal of the houses from Old London Bridge. The Act also gave a number of key directions to the commissioners. Concerning finance, it was stipulated that no tax of any sort should be levied. The level of tolls was laid down, varying from two shillings and sixpence for a coach drawn by six horses to one halfpenny for a foot passenger, or one penny if pushing a wheelbarrow. Compared with the average wage of a skilled craftsman of about 12 shillings a week, the tolls seem high, but they were similar to the tolls charged for other contemporary bridges and the ferry it was about to replace. The ferry was to be shut down on the completion of the bridge, and Henry Holland received the generous compensation of £5,350. The Act also laid down the punishment for anyone convicted of damaging the bridge. Convicts were 'liable for transportation to one of His Majesty's colonies in America for seven years'. However, the colonies decided to declare independence in 1776, a year before the completion of the bridge, so this punishment could never be handed out. There is no record of what did happen to transgressors, although, of course, Australia quickly replaced America as the normal destination for convicts.

The most controversial stipulation of the Act was the definition of the location of the bridge, which was to be 'at the Ferry or as much lower down the river as the Commission can settle'. As already stated, the inhabitants really wanted the bridge to be built at Water Lane so that access would be conveniently flat. The descent from Ferry Hill to the river was so steep that laden wagons were unable to use the ferry and had to cross further upstream via Kingston Bridge. The steepness of the incline had in fact created a business opportunity for a local woman who provided chairs for people to rest on midway up the slope. For this she was paid a few halfpence. Unfortunately, the land opposite Water Lane was owned by Henrietta, Duchess of Newcastle, the granddaughter of John Churchill, Duke of Marlborough, and she had made up her mind that she did not want the Middlesex bank approach road built anywhere near her country mansion. She proved a far more powerful and obdurate opponent than William Windham. In the end, the commissioners had to give way and agreed to start construction at the bottom of Ferry Hill, which is the site of Bridge Street today. The steepness of the slope did cause problems for users of the new bridge, and this was only alleviated in the nineteenth century, when the dip was filled in as much as possible so as to lessen the incline from 1 in 16 to 1 in 33.

Among the first decisions made by the commissioners was to choose to use Portland stone as the main construction material and to appoint James Paine as the architect, with Kenton Couse as his assistant. Strangely, there is no record of any competition for these appointments.

James Paine (1717–89)

Paine was the son of a carpenter from Andover. He trained as an architect in London, where he caught the attention of Lord Burlington, the leading proponent of the fashionable Palladian style of architecture. Burlington had built his famous Palladian villa at Chiswick, but had strong Yorkshire connections and had designed the Assembly Rooms at York. Most of Paine's commissions were in the north of England, where he designed Doncaster Mansion House and worked on the restoration of many great country houses, often in conjunction with Capability Brown, who redesigned the landscapes. He also designed a few houses in London, including Dover House in Whitehall, now the home of the Scottish Office.

With Sir Robert Taylor, Paine was considered the leading Palladian architect following the death of William Kent.

However, he had built only one bridge. This was at Shardlow over the River Trent in 1760. Couse does not appear to have had any experience with bridges and it is therefore remarkable that their combined efforts should have stood the test of time so well. After completing Richmond Bridge, Paine did go on to design three further bridges over the Thames, the last of which was at Kew, which is covered in Chapter 2.

Construction was put out to tender and a contract was signed on 16 May 1774 for Thomas Kerr to build the bridge for the sum of £10,900. It was now time to raise the money to pay him and cover all the other expenses such as building the approaches and compensating local landowners. The method chosen was known as a 'tontine', named after Lorenzo Tonti who had originated the idea in France in the 1650s. The sum of £20,000 was raised by the sale of shares which paid an initial annual dividend of 4 per cent. As each investor died, his or her share was divided between the survivors until the last survivor received the whole of the dividend, amounting to £800 per annum. When there were no more survivors, dividends would cease. In order to avoid fraud, the investors had to sign an affidavit declaring that they were still alive before they could receive the dividend, which was paid biannually. The list of shareholders held in Richmond Local History Library contains an unusually large number of investments made in the name of children. It is not therefore so surprising that the last survivor did not die until 1859, at the age of 86, having received the maximum £800 for the last five years of her life. Her identity is not known although she will have been one of the 20 investors listed as still alive in the register of 1843. Richard Crisp relates an amusing story about one of the investors, an elderly lady:

> [She] called on the paymaster, William Smith, for her
> biannual dividend and found it was the same as her
> previous one. She exclaimed in a discontented tone,

'What, has no one died since I was last here – all still alive?' But it was the last time she complained. When the dividends were next due, death had removed her, thus adding to the amount to be shared by those that survived her.[2]

Strangely enough, the shares could be sold, although the purchaser relied on the survival of the original investor to receive the dividend. In 1833, an advertisement appeared for sale to the highest bidder of a £100 share 'currently paying £14 per annum. The nominee is a lady of 69 years of age and in good health.'

Construction of Richmond Bridge started in August 1774 and the commissioners asked if the Prince of Wales would perform the ceremony of laying the first stone. Whether for lack of interest or because the Prince had another engagement, the request was turned down, so it was agreed that Henry Hobart, the leading active member of the Commission, should lay the first stone. Work progressed without notable incident apart from some complaints that the solid abutments at each end of the bridge would impede navigation and the general feeling among the commissioners that the contractor, Thomas Kerr, was proceeding too slowly. However, money was running out and a second tontine for an additional £5,000 was raised on 4 November 1776.

By the autumn of 1776, the bridge was far enough completed for foot passengers to cross. The commissioners were able to declare it open for carriages on 12 January 1777, although the contractors had yet to finish the parapets, kerbs and tollhouse. The commissioners' dissatisfaction with Thomas Kerr resulted in acrimony, especially when he demanded extra money for additional work he had done which was not in the original contract. Arbitrators were appointed and Kerr was represented by Robert Mylne, the eminent architect who had recently designed the first Blackfriars Bridge. The matter was settled when the jury, presided over by Lord Chief Justice Mansfield, decided that the sum of £89 3 s. 11 d. offered by the commissioners was fair.

The final touch to the construction work was to erect in December 1777 on the Surrey side of the bridge an obelisk, which still stands today. It is inscribed with the words, 'The first stone of this bridge was laid 23 August 1774 and the bridge was completed December 1777.' The obelisk also lists mileage to places in London, Middlesex and Surrey, including XI miles to London Bridge, X miles to Westminster and XV miles to Staines. It also warns, 'Persons who damage or deface the bridge will be prosecuted.' No mention is made of the original punishment set down in the Act of Parliament of transportation to one of the colonies in America.

Surprisingly, there was no official opening ceremony and initial public reaction was muted. However, by 1779 articles appeared in the press praising its aesthetic merits. The author of an article in the *London Magazine* of September 1779 wrote: 'it presents the spectator with one of the richest landscapes nature and art ever produced by their joint efforts, and connoisseurs in painting will instantly be reminded of some of the best

1780 view of Richmond Bridge with tollhouse on Richmond side of bridge

performances of Claude Lorraine.' This was high praise indeed, as Claude was one of the most important inspirations for the great renaissance of English landscape gardening during the eighteenth century. J.M.W. Turner admired Claude greatly and insisted in his will that his paintings should be displayed in the National Gallery only if two of his sunrises were in the same room as two paintings by Claude. In the 1820s, Turner produced about twenty sketches of the bridge from various viewpoints as well as one finished watercolour, which can be seen in Tate Britain. Constable also painted Richmond Bridge, as have innumerable other artists right up to the present day.

The bridge soon became popular with local inhabitants and also with travellers to and from London. Boswell tells the story of an evening visit by himself, Dr Johnson and Sir Joshua Reynolds to a mutual friend, Richard Owen Cambridge, who lived in a beautiful villa in Twickenham. They were travelling from London but, as Boswell wrote:

> Dr Johnson's tardiness was such, that Sir Joshua, who had an appointment at Richmond earlier in the day, was obliged to go by himself on horseback, leaving his own coach to Johnson and me. Johnson was in such good spirits that everything seemed to please him as we drove along.[3]

Reynolds himself had a house on Richmond Hill. Wick House, which still stands today, was built for him by Sir William Chambers. It was from here that he painted the famous view from Richmond Hill, but he seldom did landscapes and never painted Richmond Bridge itself. It is strange to think of the great man, the first president of the Royal Academy, galloping on horseback over Richmond Bridge while his less distinguished friends travelled in relative comfort in his coach. However, he will have saved himself money on the toll, as it cost twopence to cross the bridge on a horse, while Boswell and Johnson will have had to pay two shillings. The tolls were not at all popular. Even David Garrick, who was one of the commissioners, explained to a friend whom he had invited to dinner that he should cross the bridge by foot and a coach would then pick him up. This device was probably common for local inhabitants as the pedestrian toll was one halfpenny, compared with the coach toll of two shillings. By 1822, the bridge finances were in such a good state that all tolls were reduced to one penny. When the last survivor of the first tontine died in 1859, all tolls ceased and the tollhouses were later replaced by iron seats, dated 1868, which are still situated in the recesses of the bridge on the Richmond side. Sufficient funds were left to pay the last survivor of the second tontine until his death in 1865, nearly 90 years after the bridge was opened. Unlike at the unceremonious bridge opening, the closing of the tolls was greeted with enthusiastic shouts from the large crowds which had gathered to see the toll gates removed. Tolls were reintroduced once more for one day only on 30 May 1964 by students of St Mary's College, which is housed at Strawberry Hill, the former home of Horace Walpole. They set up a private toll by holding a long pole across the bridge and collected money from motorists for their annual Strawberry Fair charity day until the police moved them on.

Richmond Bridge with boats on the annual Trafalgar Day cavalcade

During the early years of the twentieth century, there were many arguments about how to solve the problem of increasing congestion on the bridge. Omnibuses posed a particular difficulty as they were nearly as wide as the bridge itself. The main alternatives were to build a totally new bridge or to widen the existing one. Needless to say, the inhabitants of Richmond and the press were not short of proposals, forcefully argued and equally forcefully opposed. One picture held in Richmond Local Studies Collection shows an artist's impression of how a new bridge crossing the river downstream of Richmond Bridge would entirely spoil the famous view painted by Turner. In the end, a new bridge was in fact constructed in 1933 to the north of the town to take the Chertsey arterial road over the river to Twickenham and beyond. By then, Surrey and Middlesex county councils had finally agreed that the old bridge should be widened, and its control was transferred to public ownership on 31 March 1931. The 160-year-old Commission had constructed and maintained an aesthetic masterpiece, but it was for the county councils to make it suitable for modern traffic conditions.

In 1933, Sir Harley H. Dalrymple-Hay produced a report on the condition of the bridge and how it could be widened. He suggested four alternatives. The first was to extend the footways out on either side of the bridge by supporting them on projecting stone corbels so that motor vehicles had use of the whole width of the existing bridge. The second was to extend the whole bridge equally on both sides. The third and fourth were to extend the bridge on the downstream or the upstream side only. The corbelling scheme was the cheapest, but considered aesthetically unacceptable. It clearly made sense to widen the bridge on one side only if possible, and Dalrymple-Hay concluded that the best solution was to do this on the upstream side, as this would cause the least disruption to the nearby houses. The councils approved his estimate of £73,000 and appointed the Cleveland Bridge and Engineering Co. Ltd of Darlington to run the project. Work proceeded to number each of the facing stones before taking them down so that the inner portion of the bridge structure could be widened and subsequently refaced with the original Portland stone. The result was a bridge which was widened from 24 feet 9 inches to 36 feet but looked exactly the same as before. The effect of the widening can be noted only by looking up from underneath the arches, where the newer bricks on the upstream side are clearly differentiated from the original brickwork. The work took two years and was completed in the summer of 1939 just before the outbreak of the Second World War. It caused considerable disruption, although a single line of traffic was always kept open. During the project, it was found that the original foundations, which had lasted over 160 years, were hardly state of the art. The piers were built on wooden platforms sunk only a little way beneath the river-bed, surrounded by wooden piles, some of which had rotted away. Nevertheless, the bridge was considered safe enough after sheet-steel piling was driven into the river-bed and cofferdams constructed to lay concrete foundations.

There was one further opportunity for the inhabitants of Richmond to club together in protest concerning the bridge, using what was now a permanent pressure group known as the Richmond Society. That was when the Council decided that the gaslights on the bridge would have to be converted to electricity. Various modern lamp-post designs were put forward, but the Richmond Society succeeded in forcing the Council to reject them in favour of converting the existing gas lamp-posts to electricity. There is no doubt that the old gas lamps do add to the atmosphere of times past on the eighteenth-century bridge. However, strictly speaking, they are an anachronism, as gas lighting was not invented until early in the nineteenth century, well after the bridge was completed.

The *Richmond and Twickenham Times* has reported a number of incidents involving Richmond Bridge over the past years, two of them involving potentially dangerous collisions with boats. On 20 March 1964, three pleasure boats were tied together at moorings at Eel Pie Island about one and a half miles upstream. It was a stormy night and the tide was in flood, causing all three to break their moorings and be swept down to Richmond, where

Flooded towpath by Richmond Bridge

they crashed into the bridge. No serious damage was done to the bridge and two of the boats were in good enough condition to be towed back to Eel Pie Island and repaired. However, the *Princess Beatrice*, a historic pleasure steamer built in 1896 and once used by Gilbert and Sullivan, was damaged beyond repair and had to be scrapped.

Another incident, which must have engendered a certain amount of *schadenfreude* among the less wealthy inhabitants of Richmond, occurred on the evening of 30 January 1987. The *Brave Goose*, the largest motor yacht to be built at the famous Tough Brothers' yard on the upper reaches of the Thames at Teddington, got stuck under the central arch of Richmond Bridge on its way downstream to its moorings at Tower Pier. The owner, the chairman of NCP car parks, Sir Donald Gosling, had paid £3,500,000 for the boat. The tallest part of the yacht was 25 feet high while the central arch of the bridge was 33 feet above low water mark at its highest point. Sir Donald must have spent an anxious night, and been considerably relieved when the boat was freed the next morning.

36

A more cheerful report concerned the sighting of a dolphin under the bridge on 29 October 1999. Several people reported seeing it and informed the RSPCA, who failed to locate it. However, it was later spotted swimming past the Thames Barrier, and would have completed a journey of nearly 100 miles from the North Sea up to Richmond and back. Since the Thames now contains many fish, the dolphin may well have had a pleasant journey.

Richmond Bridge is now over 200 years old. Its bicentenary was celebrated on 7 May 1977; this was exactly four months later than it should have been, but the organisers wanted to avoid the inclement weather common in January, the month in which the bridge was opened in 1777. Unfortunately, the ceremony was dampened somewhat by rain anyway. But a good show was put on by minstrels and groups in eighteenth-century dress. Today, when passing the obelisk on the Richmond side to cross the river or when viewing the elegantly proportioned arched structure from the riverside, it is not hard to imagine the scene as it would have been over 200 years ago, as Dr Johnson and Boswell paid their tolls to cross the bridge in pursuit of Sir Joshua Reynolds on horseback on their way to their friend's house in Twickenham.

Richmond Railway Bridge

Built in 1848, this was the first railway bridge to cross the River Thames in London. Up until then, the London termini served only stations on the same side of the river. The first trains had in fact come into London as early as 1836, when London Bridge Station was constructed with its short railway line to Deptford. To the west, Isambard Kingdom Brunel constructed his Great Western Railway (GWR), with its London terminus at Paddington, and the London and South Western Railway (LSWR) extended their line from their Nine Elms terminus to Southampton.

Before the coming of the railways, steam had been used from 1816 for steamboat services, which had burgeoned in a totally unregulated manner. Steamboats were much faster than the wherries and ferries rowed by watermen, which had been the favourite method of transport up until the eighteenth century, as the roads were bad and highwaymen presented a constant danger. By 1843, steamboats were running six times a day to and from Richmond. But their popularity was much shorter-lived than that of the watermen before them, as the railways soon provided an even faster and more secure service. Moreover, they were not entirely safe, as evidenced by the explosion on board the Richmond steamboat near Westminster Bridge in 1817. A cartoon was published at the time showing the steamboat exploding and its fashionably dressed passengers flying up in the air in a cloud of black smoke.

The railways received a tremendous publicity boost in 1840 when Queen Victoria herself agreed to travel from Windsor to London by train. In fact, her cavalcade drove from Windsor to Slough station, where she was greeted by Isambard Kingdom Brunel. She travelled in a specially constructed royal carriage from Slough to Paddington, accompanied by Brunel, who stood on the footplate. The journey from London to Windsor Castle, the Queen's favourite country retreat, was considerably shortened, and she expressed her approval, much to the delight of Brunel.

In 1846, a rival company to the LSWR built an extension from the LSWR station at Falcon Bridge (today's Clapham Junction) to Richmond. The opening ceremony took place on 24 July. A train took the ceremonial party from Nine Elms to Richmond in 32 minutes. According to the *Illustrated London News* of 21 October 1848, the railway 'pursues a pretty course through the villas, orchards and nursery gardens which stud that locality until it reaches Wandsworth. The River Wandle and the valley are crossed by a splendid viaduct of 23 arches.' Today's traveller between Nine Elms and Wandsworth would not recognise this idyllic description of what is now unrelenting urban sprawl. The aim of the railway was to capture most of the river traffic as well as passengers using public coaches and the new horse-driven omnibuses. The railway proved a success and by 1847, when the LSWR took over the service, it was handling 25,000 passengers a month.

The following year, the company made the momentous decision to extend the line across the river and run trains to Windsor, providing an alternative route to that taken by Queen Victoria on the GWR. Richmond Railway Bridge consisted of stone-faced land arches and two stone-faced piers supporting three 100-foot spans of cast-iron girders. Although not universally admired, the bridge was one of the more decorative railway crossings built over the Thames in London and was praised in the *Illustrated London News* of 21 October 1848 as a 'handsome structure'. The bridge was designed by Joseph Locke, the LSWR's head engineer.

Joseph Locke (1805–60)

Joseph Locke was born in Yorkshire in 1805 and worked as an apprentice to George Stephenson on the Stockton and Darlington as well as the Liverpool and Manchester railways. It has been suggested Locke was driving Stephenson's famous steam engine, the *Rocket*, when the first fatal railway accident occurred as it crushed the local MP William Huskisson on 15 September 1830. If so, it did not deter him from working on many other railway projects, including the world's first long-distance railway, linking Birmingham to the Liverpool and Manchester Railway, in 1837. Most of his early railway commissions were in the north of England and Scotland. He later joined the London and Southampton Railway, which became the London and South Western Railway, for whom he designed the Richmond Railway Bridge and also Barnes Railway Bridge. During his time at the LSWR, he stood for Parliament and was elected Liberal MP for Honiton in Devon. In 1857, he was elected president of the Institute of Civil Engineers. Together with Robert Stephenson and Isambard Kingdom Brunel, Locke is considered to be one of the great triumvirate of Victorian civil engineers who pioneered the Railway Age in Great Britain. Unlike the more famous Brunel and Stephenson, he left behind no spectacular monuments, but many believe he exerted a greater long-term influence on railway engineering than either of them.

One predictable result of the new rail link to Windsor was an increase in demand for visits to the castle. To avoid inconvenience to Queen Victoria, the State Rooms themselves were open only at specified times. Entry tickets were free at the Queen's insistence, while guidebooks cost one penny.

The cast-iron girders of the bridge were replaced by steel ones in 1907 with little change to its appearance. The bridge, like all the others over the Thames, is inspected by divers about every five years. It is still in good condition, despite having to survive the impact of a 120-ton barge, which crashed into it at 3 p.m. on 28 December 1996. Part of the barge was ripped off. However, Railtrack reported that 'there was only superficial damage to the bridge and the line was opened at 4.05 p.m. These Victorian structures really are pretty solid.'

Richmond Railway Bridge

Twickenham Bridge

Twickenham Bridge was completed in 1933 as one of the three bridges needed to take the Chertsey arterial road across the Thames to improve the traffic flow from London to the Portsmouth road and thereby relieve increasing congestion on Richmond Bridge, which had not yet been widened. The other bridges built at the same time were Chiswick Bridge and a replacement Hampton Court Bridge. The scheme had first been put forward in 1909 but, because of the 1914–18 war and arguments between the various interested groups about the exact route and financing, it was not until the report of the Royal Commission on Cross-river Traffic of 1926 that the Ministry of Transport (MOT) decided to go ahead. The MOT agreed to contribute 75 per cent of the cost of the bridges and their approaches. Middlesex and Surrey county councils then put forward a Bill and Royal Assent was given on 3 August 1928.

Local inhabitants and Richmond Council were incensed to find that the MOT had agreed without consulting them to redirect the road from the original route, which went through some disused land. Evidently the Board of Trade had sanctioned the erection of an experimental plant by the Gas Light & Coke Co. on this land. This meant that houses and shops on the Lower Richmond Road would have to be demolished. Questions were raised in Parliament, but it seems that national interest overruled local opinion. Eventually, over 300 families had to be rehoused in new blocks of flats or houses.

Maxwell Ayrton (1874–1960) was appointed architect for the project, and Alfred Dryland (1865–1946) was appointed head engineer. Ayrton originally designed an imposing structure with four 70-foot towers at the riverbanks and flanking walls 20 feet above road level. The *Daily Telegraph* conducted a campaign, with input from locals, against this proposal, on the grounds that the design was inappropriate for the quiet river setting of Richmond. The matter was referred to the Royal Fine Art Commission, which included, among other eminent people, Sir Edwin Lutyens (1869–1944), who designed Hampton Court Bridge, and Sir Reginald Blomfield (1856–1942), who designed

41

Lambeth Bridge. The commissioners agreed that the design was inappropriate for this reach of the river, however fine it might have looked in more dramatic scenery. 'Its dramatic and fortified appearance seems so foreign to its quiet surroundings and presents so striking a contrast to the neighbourhood' that they felt the architect should be asked to simplify his design. These opinions were more forcefully expressed by the locals. Mr M. William Jones wrote to the *Daily Telegraph*:

> May I thank you for your invaluable support of those who are trying to save the scenery from the latest scheme to ruin its beauties? The old railway bridge is an abomination; the lock bridge close at hand is as bad; the proposed new bridge will be ten times worse than either. Its site is some 300 yards downstream from our old bridge, and its blatant ugliness will wreck all that is left of the beauty of this reach.

Another resident proposed that the new bridge should be built over and on each side of the railway bridge so as to minimise the number of bridge structures in the area. In the end, Ayrton agreed to change the design. At first, he omitted the towers on the Middlesex bank but kept shorter towers on the Surrey bank; the final design, however, has no towers and was highly praised at the time. *Country Life* published a photograph of the new bridge on 8 July 1933 with the caption 'Twickenham Bridge. A beautiful concrete structure'.

The original name of the bridge as specified in the Act of Parliament was Richmond Bridge, but this was confusing, as the old bridge was also called Richmond Bridge. Some proposed that it should be called Queen Elizabeth's Bridge to commemorate the fact that the great queen died at Richmond Palace, which fronted the river just by the Surrey end of the bridge. However, in the end, the more mundane name Twickenham Bridge was chosen.

The contract for the construction of Twickenham Bridge, which stretched 2,500 feet including the approaches, was

Twickenham Road Bridge viewed through the Railway Bridge riverside arch

awarded on 1 June 1931 to Aubrey Wilson Ltd for the sum of £191,206. The 70-foot-wide bridge was built to cross the Thames 200 feet downstream of the railway bridge, where it crosses the 280-foot width of the river in three reinforced concrete spans. Between 175 and 200 men worked on the construction of the bridge, which was completed on time and opened on 3 July 1933 by HRH Edward, the Prince of Wales. The river arches have three permanent hinges, Twickenham Bridge being the first reinforced-concrete bridge in the UK to use this principle. This method leaves the arches free to adjust themselves to changes in temperature. The hinged arches of the river spans are emphasised by the provision of bronze hinge plates at the springings and centres. The distinctive shapes in pre-cast concrete placed within arches above the cutwaters give the design a definite art deco flavour.

The later history of Twickenham Bridge was uneventful until major repairs were undertaken in 1994. The *Evening Standard* of 28 October 1994 ran a banner headline, 'Chaos ahead as bridges are fixed'. The work was planned to take five months and major disruption to London's traffic was anticipated. The report commented that bridge closures are always a headache for London's traffic, not least because so many different bodies are involved, including 11 borough councils and the Department of Transport. Today, we would have to add the Greater London Authority (GLA), although this is supposed to improve matters. With 750,000 vehicles crossing 23 bridges each working day, the closure of a single bridge is sure to cause chaos.

Richmond Footbridge, Lock and Weir

The year 1894 saw the opening of the two most idiosyncratic bridges over the Thames in London: Tower Bridge and Richmond Footbridge, Lock and Weir. Tower Bridge is world renowned; few people know about Richmond Footbridge unless they live nearby. However, *The Times* of 16 May 1894, reporting on the opening ceremony, states:

> The ceremony will complete one of the most memorable events in the history of the river, perhaps second only in importance to the demolition of Old London Bridge, for even the construction of Teddington Lock had not the same far-reaching influence as will be exercised by this newer structure.

It was the weir which excited this eulogy from *The Times* rather than the footbridge, which, as we will see, was really an afterthought, designed to improve the appearance of the structure.

In the latter half of the nineteenth century, there was an increasing problem with the lack of water depth in the river between Teddington and Richmond. There were two major reasons for this. First, the water companies were removing more and more water from the Thames at Staines to fill the reservoirs which provided a clean water supply to Londoners. Second, the removal of Old London Bridge, with its 19 arches, had allowed the tide to ebb and flow much faster, as the replacement bridge built by John Rennie, with five much wider arches, no longer acted as a sort of dam. Consequently, the flood tide poured up the river in a great rush as far as Teddington and then, after a short period of slack, the tide would retreat equally quickly and carry nearly the whole volume of water back towards London and the sea. The result was that at low tide the stretch of the river around Richmond turned into a trickling ditch in the middle of a mudbank. In addition, according to *The Times*, 'the scour of the tides brought up all the filth from the sewage outfalls below London, and this, deposited on the foreshore, produced huge banks of mud where it had been almost as clean as a beach'.

The effect on the local environment and economy was disastrous. Apart from the despoliation of the scenery, riverside house prices fell to such an extent that one house bought for £14,000 in 1877 was sold for only £10,000 in 1889. Fish stocks dwindled and large fish such as salmon and trout vanished altogether, to the dismay of the 30–40,000 local fishermen. The watermen who took passengers up and down the Thames also lost business, as their boats could operate only for a few hours a day. A group of them even tried to bring attention to their plight by playing a game of cricket on the river-bed at Twickenham at low tide.

The first protest was delivered as early as 1860 to the Thames Conservancy, which was responsible for the state of the Thames at that time. Many further protests led to the conservators agreeing to dredge the appropriate stretch of the river at a cost of £21,000. This turned out to be a waste of money. The solution proposed by the local authorities, the Richmond Vestry and Twickenham Local Board, was the construction of a weir and lock. However, this continued to be resisted by the conservators, so eventually the local authorities decided to lodge a Bill in Parliament and, despite further opposition, an Act was passed in 1890 'to authorise the construction of a footbridge with removable sluices and a lock and slipway on the River Thames in the parishes of Richmond and Isleworth'. The estimated cost was £61,000, of which £40,000 was raised by local rates and the remaining £21,000 from the resources of the Thames Conservancy, which was made responsible for the construction project. Tolls were also introduced for the footbridge, to offset the contribution of the Thames Conservancy and to cover maintenance. A strange anomaly arose when it was found that the toll of one penny to cross the footbridge meant that anyone wanting to go onto the bridge just for sightseeing would have to pay twopence as they had to pass the toll a second time to return to the riverside. The tolls were not abolished until after the Second World War.

The design of the combined weir, lock and footbridge required remarkable ingenuity. The main problem was how to construct a moveable barrage which had to operate twice a day to control the great volumes of water flowing up and down such a major river as the Thames. Such barrages had been built on several rivers in Europe but they were designed to control the flow of the river only twice a year at the start and end of the rainy season and were far too cumbersome to be opened and closed twice daily. For the recently constructed Manchester Ship Canal, Mr F.G.M. Stoney (1837–97) had built over 100 moveable sluice gates which could be raised and lowered in minutes to control the level of the waters of the rivers feeding the canal. The problem with Stoney's sluices was that the sight of the massive

Richmond Footbridge, Lock and Weir

iron gates raised above the Thames at Richmond would ruin the beautiful riverside scenery. The solution was to build a double footbridge above the sluice gates and to provide a mechanism which turned the gates from vertical to horizontal so that they were hidden underneath the footbridge when raised to full height.

The structure as built consisted of a lock, weir, slipway and double footbridge. The lock was the largest on the Thames, capable of accommodating a tug and six barges. This was designed to provide a bypass of the weir so as to allow continuous passage for the 30 to 40 barges that were towed each day to and from Brunel's Brentford Dock at the time. The weir consisted of sluice gates which were raised out of the river to allow the free flow of water until, on the ebb tide, the water had fallen to the half-tide level. At this point, the gates were lowered into the water to ensure it was maintained at the half-tide level until they could be raised again when the water level had been restored by the incoming tide. This solved the problem of the emptying of the river between Richmond and Teddington during low tide and allowed unimpeded passage of shipping underneath the gates at high tide. The sluices could be raised by a hand winch in seven minutes. Today, the operation has been mechanised. The slipway consisted of a set of rollers, allowing pleasure boats to pass without needing to use the locks when the sluices were in operation. The bridge itself was designed to conceal the ungainly sight of the raised sluice gates and to match the look of the relatively decorative Victorian iron railway bridge situated 200 yards upstream.

After a three-year construction project, the combined structure was opened on 19 May 1894 by Prince George, then Duke of York and later King George V. The ceremony consisted of the royal procession accompanied by military bands and aquatic sports, and followed by an evening illuminated river fête and fireworks. In 1908, the Port of London Authority (PLA) took over all responsibility for the tidal Thames, including the Richmond Footbridge, Lock and Weir, from the Thames Conservancy. By 1970, the PLA was virtually bankrupt following the closure of the London Docks and even tried to hand over the structure, which had fallen into disrepair, to the local authorities who were responsible for the other bridges in the area. Eventually, following years of local campaigning, the PLA agreed to renovate it at a total cost of £4,000,000. In 1994, centenary celebrations were held, attended by the present Duke of York. The crowds enjoyed a vintage boat rally and traditional refreshments were provided by a beer tent and the roasting of an ox. The centenary programme describing the structure stated that 'the method by which it operates is ingenious Victorian engineering which has stood the test of time' and added that the process of renovation would ensure its survival for another 100 years.

CHAPTER 2

Kew

Kew today has two bridges, the road bridge of 1903 and the railway bridge of 1869. The road bridge is the third to be built on this site, and it replaced a much admired stone structure by James Paine, the architect of Richmond Bridge. The current bridges cater for the thousands of road and rail passengers who cross the river here each day. Although, unlike Richmond Bridge, neither can claim to be an aesthetic masterpiece, attempts have been made to add a degree of ornamentation to what are basically functional designs.

Kew Bridge

The original Kew Bridge of 1759 was the fourth bridge to be built over the tidal Thames, predating Richmond Bridge by 17 years. The village of Kew grew more or less in parallel with Richmond from medieval times, as it was conveniently near Richmond Palace and so attracted royal and aristocratic members of the court. In fact, Kew's royal connection lasted longer than Richmond's, and resulted in the creation of the Royal Botanic Gardens, which had an immense effect on the development of the village and the consequent demand for a permanent river crossing. The name Kew comes from a Saxon word meaning 'quay', which indicates the village's early origins and the primary local industry of fishing. On the other side of Kew Bridge is Brentford, which is at least as old as Kew. The name suggests that the river was fordable here at one time, and many historians believe this was where Julius Caesar crossed the Thames during his invasion of 55 BC. Unlike Kew, Brentford has had hardly any connections with royalty, although once, noticing its dirty, ill-paved streets, George II is said to have remarked, 'I like to ride through Brentford, it is so like Hanover.'

Until the eighteenth century, neither Kew nor Brentford had large populations, except during the intermittent residence of the court at Kew. The two main local industries were fishing and the ferry. Fishing lasted until the river became polluted in the nineteenth century. The industry is commemorated by names of roads near Kew Bridge, such as Westerly Ware, which refers to a weir installed to trap fish, and Old Dock Close, where the fishermen kept their boats. The ferry, as at Richmond, belonged to the monarch, who granted the monopoly lease to royal favourites. Known as the King's Ferry, it ran between the present riverside car park of Kew Gardens and a point just downstream of where the Grand Union Canal enters the Thames at Brentford.

Only in the eighteenth century did Kew grow from a collection of royal mansions and fishermen's cottages into a typical English village. The main impetus was the creation and growing popularity of Kew Gardens. In 1731 occurred possibly the most significant event in Kew's history when Prince Frederick, the elder son of George II, acquired a mansion known as the White House, situated near Kew Palace inside today's Royal Botanical Gardens. Prince Frederick is known now, if at all, as 'poor Fred', from the rather cruel Jacobite poem which starts:

> Here lies poor Fred,
> Who was alive and is dead.

However, it is likely that Kew would never have had the world-famous Royal Botanic Gardens without him, and the whole history of Kew, including the bridge, would have been different. Frederick was a man of many enthusiasms including, strangely for a man brought up in Hanover, the game of cricket. He once captained a team playing a match against a side captained by the Duke of Marlborough on Kew Green. His greatest enthusiasm was for gardening. This he shared with his wife, Augusta. Sadly, he died in 1751 before his plans for the Botanic Gardens reached fruition, but Augusta was determined to carry on the work. It is possible that she was the driving force behind the project, as, according to E.B. Chancellor, 'from her proceeded the scientific impress which has given it its unique place among national possessions, and its supreme rank among the botanical institutions of the world'.[4]

Meanwhile, wealthy people started to build the attractive Georgian houses around Kew Green which characterise the area today. St Anne's Church, built as a small chapel in 1710, was extended. Writers and artists were also attracted to the village, including Thomas Gainsborough, Johan Zoffany and Jeremiah Meyer.

The result was increased use of the ferry. The records show that Prince Frederick himself was one of the most frequent users of the service. The receipts book for 1732 to 1737 shows that the total income from tolls for the five years was £3,628 5 s. 6 d. Prince Frederick himself took 116 horses over the river by the ferry in 1736 and paid a total of £2 2 s. 10 d. However, as was the case at Richmond, the ferry was not without danger. It seems likely that the ferrymen were not always entirely sober, as there was a public house on the Brentford side, and people and horses ended up in the river on several occasions. The inn was also frequented by highwaymen, who could escape across the river after robbing people on the streets of Brentford.

In fact, by the eighteenth century, there were two ferries between Kew and Brentford: one was the above-mentioned King's Ferry and the other was a foot ferry which crossed the river along the line of the present bridge. A Brentford businessman, Robert Tunstall, had bought up both ferries and in 1757 saw an opportunity to profit from building a permanent river crossing without having to compensate any ferry owners. This may not have been the first attempt at a dry crossing if local rumours are to be believed. It is said that Oliver Cromwell escaped a party of Cavaliers after enjoying the hospitality of the Bull's Head, which is on the Middlesex bank near today's railway bridge, by means of a tunnel which led to the island in the middle of the river known as Oliver's Ait. However, even if this is true, there is no record that the tunnel continued on to the Surrey bank.

Tunstall presented a Bill in Parliament for the building of a wooden bridge along the route of the King's Ferry. This was passed, and the Act received Royal Assent on 28 June 1757. The inhabitants were clearly agreed on the need for a bridge and, unlike at Richmond, raised no strong objections to its being made of wood. They did, however, present a petition that the site of the bridge should be changed to where the foot ferry operated, as this would provide less disruption to navigation. Consequently, a new Bill was introduced and the subsequent Act received Royal Assent on 23 March 1758. This was a relatively speedy process compared with the multiple delays met by most bridge proposals to Parliament. The resulting construction project was also amazingly quick and the bridge was completed in just over one year. John Barnard, who had been the master carpenter for Westminster Bridge, was chosen to design the bridge, which consisted of eleven arches. The two piers and their arches at either end were built of brick and stone, while the other seven arches were built entirely of wood.

The year in which the bridge was opened, 1759, was also the year of the public openings of both Kew Gardens and the British Museum. Frederick's elder son, Prince George, conducted the bridge's opening ceremony on 4 June, which happened to be his birthday. George's mother, the Dowager Princess Augusta, who had devoted so much of her time to Kew, accompanied him. It seems fitting that her name as well as his appeared on the dedication inscribed on the original design of the bridge.

Prince George is known to have used the bridge frequently after it opened, as he often visited his childhood home, the

Robert Tunstall's wooden Kew Bridge of 1759, viewed from the north

White House. On 25 October 1760, as he was riding across the bridge from the White House, he was met by a messenger from London who brought the news of the death of his grandfather, George II, and hence his own accession to the throne. Horace Walpole records the event as follows:

> Without surprise or emotion, without dropping a word that indicated what had happened, he said his horse was lame and turned back to Kew. At dismounting, he said to the groom, 'I have said that this horse is lame; I forbid you to say to the contrary.'[5]

This cool-headed reaction from the man we now know as mad King George III may seem surprising. However, it is well known that George II and his elder son, Prince Frederick, who predeceased him, felt a strong mutual hatred towards each other, and Prince George is unlikely to have felt any sadness at the death of his grandfather.

The relationship between George III and his son, the future George IV, was as bad as that between George II and Prince

Frederick. Mrs Papendiek, a lady-in-waiting to Queen Charlotte, provides a fascinating insight into the life of the court in Kew at this time in her diaries.[6] She directs her venom at George III's son, who showed no concern for his father when he suffered from his spells of madness. She records the Prince's disappointment when the King recovered from his first illness just before the Lords were to give assent to the first Regency Bill which would have made him Prince Regent some years before this actually happened. She also records one occasion when she and her party were crossing Kew Bridge in 1780 during the Gordon Riots. She noted fires coming from no fewer than 11 places in London. She does not mention the payment of any toll on this occasion.

Robert Tunstall had provided the finance for the bridge himself and hoped to make a large profit from tolls. These were set as follows:

> For every coach drawn by six or more horses, 2 s.
> For the same drawn by two horses, 1 s.
> Drawn by one horse, 8 d.
> For every baker's cart drawn by one horse, 6 d.
> For every led horse or ass, 2 d.
> For every foot passenger, ½ d.[7]

At first, he must have thought he had made a profitable investment. On the first day alone, 3,000 people crossed the bridge. Robert Tunstall's initial success, however, proved short-lived, as the costs of maintaining a wooden bridge over the fast-flowing Thames soon ate into any profits he made from the tolls. The bridge did outlast him, but his son, who was also called Robert, decided in 1782 that he could no longer afford to maintain the old bridge and presented another Bill to Parliament to allow him to pull down the wooden bridge and build a new one with seven arches made of stone. *The Diary; or Woodfall's Register* of 24 September 1789 commented on the removal of the wooden bridge at Kew: 'There will be no more wooden bridges over the Thames. Those of Battersea, Putney

&c., which are hastily decaying, will be rebuilt at some future period with stone in every way more beautiful and lasting.' This prediction proved only partially correct, as the writer had ignored the possibilities of an even stronger although perhaps less beautiful material – iron. Abraham Darby had just constructed his Iron Bridge over the Severn Valley at Coalbrookdale, in 1779, but this pioneering project had no successors until the beginning of the next century.

The second Robert Tunstall had formed a business partnership for the rebuilding of the bridge with a carpenter, Charles Brown, and with his brother-in-law, John Haverfield. Haverfield was the son of the chief gardener at Kew Gardens, who had been appointed by Augusta in 1759, making the connection between Kew Bridge and the Botanic Gardens now even closer. The estimated cost of £16,500 was raised by a tontine, as at Richmond. Again as at Richmond, James Paine was the architect.

Work started on 4 June 1783 and the project lasted six years. *The Gentleman's Magazine* of November 1789 reported that the project was completed with no loss of life, which is remarkable considering the circumstances of the time and the length of the project. One advantage for the builders was that the old bridge was retained about 100 yards upstream until the new bridge was completed, thus obviating the need for putting up a temporary structure. Various cost-saving measures were introduced as the estimate of £16,500 proved too low. Brick was used instead of stone for the land arches and the foundations were not dug deep enough. Nevertheless the architecture of the bridge was admired almost as much as that of Richmond Bridge, and J.M.W. Turner, who had lived in Brentford as a young boy, painted a watercolour of it which is now in Tate Britain. This shows the steep curve of the bridge as it arches over the river. The curve gave it a picturesque appearance but was much criticised by the people who used it.

The new stone Kew Bridge was finally opened by George III on 22 September 1789, almost exactly 30 years after he had opened the first bridge as Prince of Wales. The storming of the

James Paine's stone Kew Bridge of 1789

Bastille had occurred in July of that year and George III had already survived an assassination attempt shortly before this, but he always felt at home in Kew and clearly wanted to be associated with its opening. The King insisted it should not be opened until he had crossed it himself and even proposed to purchase it and relieve the public of having to pay tolls. This never happened, though, and the tolls continued to be collected. After the opening ceremony, a celebration dinner was held at the Star and Garter but James Paine was too ill to attend and sadly died soon afterwards.

Unlike the first bridge, the second proved very profitable for Robert Tunstall, as usage increased with the growing popularity of Kew Gardens and maintenance costs did not escalate so dramatically. Even royalty were not exempt from the tolls, although it is recorded that both Queen Charlotte and the Prince Regent ran up considerable debts. However, in 1824, Tunstall sold the bridge to a Mr Thomas Robinson for £20,000. This may well have been because he feared that the projected suspension bridge at Hammersmith, which was to be opened in 1827, would divert traffic from Kew Bridge, as it provided a more direct route to London via the new road through Barnes.

Had he lived until 1874, Robert Tunstall might have come to regret his decision to sell the bridge, because in that year a joint committee of the Metropolitan Board of Works (MBW) and the Corporation of London purchased it from the new owners for £57,300. This happened after many years of public pressure to free London's Thames bridges from tolls, which had become almost as unpopular as the Poll Tax. There were several reports of attacks on toll-keepers and often the attackers were let off lightly. However, on one occasion, after a fight between the toll-keeper on Kew Bridge and a cabman called Thomas Johnson, it was the toll-keeper who came off better. When Johnson died from his injuries in St Bartholomew's Hospital, the toll-keeper was tried for manslaughter but acquitted by the jury.

Matters came to a head in 1868 when a government Bill was presented to allow duties on coal and wine to continue until 1889. These duties were supposed to end in 1869, but the Government hoped to earmark them to fund the Thames Embankment. Strenuous opposition resulted in a compromise that allowed the duties to continue on the condition that a clause was inserted into the Bill which stipulated that the revenues should first be applied to the freeing from toll of Kew, Kingston, Staines, Walton and Hampton bridges. In 1869, an Act was passed to empower the MBW and the Corporation of London to purchase all these bridges and remove all tolls from them. Tolls on Richmond Bridge had already been ended by the commissioners, as mentioned in Chapter 1. The days of privately owned toll bridges were numbered.

Staines, Walton and Kingston bridges removed tolls soon after the Act of 1869, but the freeing of Kew Bridge from tolls was delayed because the proprietors demanded compensation of £70,000, which was far more than the £39,000 estimated by the county councils. In the end, a compromise was reached, and at the ceremony commemorating the removal of the tolls, the Lord Mayor first asked the proprietors if they had received the agreed sum of £57,300. When they answered in the affirmative, the Lord Mayor unlocked the toll-gates to rounds of cheering from the massed crowds and the booming of cannon. Firemen rushed forward, raised the gates from their hinges and bore them in triumph to a brewer's dray drawn by two white horses and driven by a man in a red cap. The Lord Mayor's party proceeded off the bridge and round Kew Green and ended back at the Star and Garter on the north end of the bridge for luncheon. The watchword in the surrounding districts had been 'Free bridges for a free people', and this was displayed on banners all around the area. On a more trivial note, the tolls had provoked a series of jokes, including, 'Why is the toll payable at Kew Bridge like a clergyman's deputy? Because it is a Kew rate [curate].'

The year 1869 was a significant one for the inhabitants of Kew. Not only did it see the Act passed to remove the tolls on Kew Bridge, but it was also the year when the railways came to Kew. This required the construction of a railway bridge over the Thames, described later in this chapter. The result was to change Kew from a riverside village into a suburb of London and to increase the accessibility of Kew Gardens to such an extent that by the 1880s over a million visitors a year were recorded. As Charles Dickens wrote in his *Dictionary of the Thames*, 'Kew is losing most of its distinctive features. But for the quaint old Green with its picturesque surroundings, there is little to remind us of the Kew of even 20 years ago.' By the 1890s, with the rise in population and following the abolition of tolls, Kew Bridge could no longer handle the heavy increase of traffic. It was too narrow for new modes of transport such as horse-drawn omnibuses and the steep ascent to the middle made the crossing difficult and dangerous for heavy vehicles. The *Daily Graphic* of 25 July 1896 described the frequently chaotic traffic conditions on the bridge: 'Heavy drays and wagons charge down from the high road by Kew Bridge Station in order to gather sufficient momentum to carry the struggling horses to the crown of the bridge.' The writer expressed amazement that no serious accident had yet occurred.

Middlesex County Council was in favour of rebuilding the bridge in stone, but Surrey County Council, which was jointly responsible, was not convinced that a new bridge was needed and anyway preferred an iron bridge because it did not want to incur the extra expense a stone one would involve. Eventually, Middlesex obtained agreement that they would pay for the cost of a report to establish whether the existing bridge was safe and which material should be used if a new one was required. Sir John Wolfe Barry, who had done the engineering design for Tower Bridge, was asked to report on the state of the bridge with regard to both traffic conditions and structural safety, and to recommend which type of bridge should be built. His report of 1892 concluded that the bridge was not only inconvenient but also in a dangerous condition, not least because of inadequate foundations. He also recommended that a new bridge should be constructed in stone. Sad as it would be to lose Paine's picturesque stone structure, the only solution was to pull it down and erect a new bridge.

The Kew Bridge Act received Royal Assent on 25 July 1898, empowering Surrey and Middlesex county councils to rebuild Kew Bridge and make the necessary new approaches. The cost was £250,000, paid for out of council taxation. John Wolfe Barry and Cuthbert A. Brereton were appointed engineers, and the contract was awarded to Mr Easton Gibbs of Skipton following a competitive tender. The project lasted five years, during which a temporary wooden bridge was used to provide for a limited amount of traffic to cross the river. The old bridge was removed in an ingenious manner. A cableway technically known as a Blondin was stretched across the river. This allowed the workmen to remove large chunks of the bridge and carry them away at the rate of 750 cubic feet a minute, which was much faster than was possible with the use of barges.

The new bridge was constructed mainly of Cornish granite. It consisted of three spans and had a 56-foot-wide roadway, compared with the seven spans and 18-foot-wide roadway of Paine's bridge. While the piers of the old bridge were supported on wooden platforms which were sunk only a few feet below the river-bed, the foundations of the new bridge were carried down 18 feet into the London clay under the river. Viewed from the river, it is a handsome structure with its three remarkably flat elliptical arches. The solid roughness of the granite is relieved by several ornamental elements, including four shields bearing the arms of Middlesex and Surrey carved into the walls. The *Art Journal* of 1899 was ambivalent and regretted the passing of the old stone bridge. Writing at the start of the construction project, it stated that the new granite structure struck a false note in the traditional riverside landscape where

> the quaint old houses of Strand on the Green will still be standing and groups of barges and lighters will still cluster at the north end, and busy carts ply to and from them as they lie high and dry when the tide is down. The workers in the market gardens, the market carts with their gaily painted chamfering will cross and re-cross as before.

The writer would have been even more horrified if he could have imagined modern traffic over the bridge. However, he did admit that the new bridge was more convenient for the visitors who crossed the river on their way to Kew Gardens and was less trying for the horses.

The opening ceremony on 20 May 1903 was conducted by Edward VII and Queen Alexandra, who had recently been crowned following the death of Queen Victoria in 1901. In deference to the King, the bridge was named the Edward VII Bridge, although it has always been known as Kew Bridge, just like its predecessors. The King and Queen went in procession from Buckingham Palace to Kew through London's western suburbs of Kensington, Hammersmith, Chiswick and Brentford. As *The Times* reported:

> enthusiastic crowds lined the route from the Palace as the King and Queen drove in an open landau drawn by four horses. All along the route it seems people decked their houses and streets with flags and banners with such loyal

greetings as 'Our worthy King, God Bless him. In thy right hand carry gentle peace.'

The trowel used by the King at the ceremony was made out of wood from the old wooden bridge, which had by now become a collector's item. A number of gifts were presented, including a prehistoric axe found during the excavations for the foundations of the new piers and a chair with representations of the three Kew Bridges carved into the ladders in its back. The whereabouts of these gifts today are not known, but a replica of the chair can be seen at Middlesex Old Guildhall in Parliament Square.

As far as artistic representations of Kew Bridge are concerned, although there are several engravings, mainly of the second bridge, and a watercolour by J.M.W. Turner, it has not received the attention accorded to Richmond Bridge, which is set in a more dramatically beautiful stretch of the river. It has, however, made a few appearances in literature. Charles Dickens' *Oliver Twist* crossed the bridge in the company of Bill Sikes on his way to the bungled robbery in Hampton. It also features in Jerome K. Jerome's *Three Men in a Boat* when one of the heroes, George,

The solid granite structure of John Wolfe Barry's Kew Bridge, opened in 1903

first tries his hand at rowing in an eight-oared boat at Kew. After a disastrous start:

> They passed under Kew Bridge, broadside, at the rate of eight miles an hour . . . How they got back George never knew, but it took them just 40 minutes. A dense crowd watched the entertainment from Kew Bridge with much interest, and everybody shouted out to them different directions. Three times they managed to get the boat back through the arch, and three times they were carried under it again, and every time 'cox' looked up and saw the bridge above him he broke out into renewed sobs. George said he little thought that afternoon that he should ever come to like boating.

The book was written in 1889, when much of the traffic across James Paine's stone bridge would have been pedestrian.

Most of the traffic over the 1903 bridge is now motorised, so reckless rowers would not receive the same attention today. Although designed for horse-drawn vehicles, it has proved perfectly suitable for modern traffic and is in good shape after more than 100 years. Kew Bridge had been high on the German hit list during the Second World War, as shown on a map found by the Allied Forces. Bombs did cause splintering on the stonework, but the bridge survived and the centenary was celebrated on 20 May 2003. When the bridge was built in 1903, many local inhabitants were concerned that it would transform the character of Kew, which prided itself on remaining a village, something it had achieved partly due to the rural appearance of the old stone bridge. All agreed that it would bring enormous improvements in communications to the north but it would also bring Kew into the mainstream of metropolitan growth. Although all this has happened, the area around Kew Green, with its Georgian houses, cricket pitch, pubs and St Anne's Church, still conjures up an old-fashioned village atmosphere both for local people and visitors who cross the Green to the ornate wrought-iron entrance gates to Kew Gardens.

Kew Railway Bridge

Whereas Kew had pre-empted Richmond by 20 years in building a road bridge, it was 20 years later in constructing a railway bridge. Because of the bend in the River Thames, Kew was too far north for the London and South Western Railway to include it on its route from Nine Elms to Richmond in 1848. However, by 1862 the company decided to provide a service by building a branch line from South Acton, which was on the railway on the north side of the Thames, via Kew Gardens Station to Richmond. Kew Gardens Station opened in 1869 and still retains many of the original decorative cast-iron columns. In 1877, the Metropolitan Railway extended its Underground service over the bridge to Kew Gardens and Richmond.

The necessary railway bridge was designed by W.R. Galbraith (1829–1914). Galbraith was born in Stirling and worked on railways in Scotland and north-west England before joining the LSWR in 1855 as chief engineer. He was responsible for a massive expansion of services to the south-west of England as far as Devon and Cornwall. In the 1880s, he went on to construct the railway lines continuing north from the newly constructed Forth Bridge, and he continued working well into his 70s while building 14 miles of Tube lines for the London Underground, including the line known as The Drain, which linked Waterloo to the City.

The Kew Railway Bridge took five years to build and was opened without ceremony in 1869. Locals often call it Strand Bridge to distinguish it from the main Kew Bridge just upstream. The name comes from the stretch of the river on the north bank to the west of Kew Bridge which is known as Strand on the Green. The bridge consists of five 115-foot spans of wrought-iron lattice girders supported on cast-iron columns with ornate capitals. The riverside abutments are of brick with ornamental stone mouldings. Train passengers will not be aware of the architectural merits of the bridge, but will have a fine view to the west of Kew Bridge and the looming tower of Kew Bridge Steam Museum beyond. Walkers and cyclists along the Thames Path can have a close look at the structure of the bridge, although they are more likely to be heading for the two pubs between which the bridge crosses the river. There they will doubtless be regaled with stories of how, in the nineteenth century, the Lord Mayor used to moor his ceremonial barge at the City Barge pub in winter, or how Oliver Cromwell escaped from the Bull's Head under the river to Oliver's Ait. The one disadvantage of this idyllic riverside setting is that the stories will be regularly interrupted by the clatter of trains crossing the railway bridge overhead.

As reported in the *Richmond and Twickenham Times* of 26 September 1986, an attempt was made to brighten up the dull grey structure of the bridge by painting it in bright colours. Workmen were photographed on platforms underneath the bridge, with straw bales hanging down from one of the navigational spans. This is the traditional method of warning shipping that work is going on overhead. The article also mentions that Sir Donald Gosling's motor yacht, the *Brave Goose*, was scheduled to pass through the bridge during the painting project and there was concern that it might have trouble because of its height out of the water. In fact, it was at Richmond

Kew Railway Bridge with the City Barge pub in the background

Bridge that the boat got stuck, as described in Chapter 1. Today, the bridge has returned to the traditional battleship-grey colour.

In 1998, Railtrack requested planning permission from Hounslow Council to repair the 130-year-old bridge, as it was in a dangerous condition. The engineers had found that the timbers supporting the rails were rotting in places and had a life expectancy of less than two years. The wrought-iron cross girders had corroded in several places where they were in contact with the timber. The solution was to replace all the timber supports with steel and in addition to renew any parts of the girders that had become corroded. Now that the repairs have been completed without any change to the external appearance of the bridge, passengers can feel safe as they cross the river on the District or North London Line and admire the riverside views. Unfortunately, customers in the historic pubs below still undergo the same noisy disturbance every time a train passes.

CHAPTER 3

Chiswick and Barnes

Chiswick Bridge

As described in Chapter 1, the 1926 Royal Commission on Cross-river Traffic had recommended that three bridges over the Thames at Chiswick, Twickenham and Hampton should be constructed to take the Chertsey arterial road from London to the west. An Act of 1928 empowered the Middlesex and Surrey county councils to carry out the work, helped by a contribution of 75 per cent of the cost from the Ministry of Transport.

The Councils appointed Alfred Dryland as engineer both for Chiswick Bridge and Twickenham Bridge. Sir Herbert Baker (1862–1946) was appointed as architect of Chiswick Bridge. After the design had been approved, the contract was awarded to Cleveland Bridge and Engineering Co. for the sum of £208,284. For once, the project was completed on time and within budget, and the bridge was opened in 1933 by HRH Edward, Prince of Wales, on the same day as the other two bridges.

Chiswick Bridge crosses the Thames about 100 yards above the winning post of the Oxford and Cambridge Boat Race, where the river is 450 feet wide. The bridge comprises three arched ferro-concrete spans, the centre one being 150 feet wide, which was the longest concrete span on any Thames

Chiswick Bridge with rowers training near the Boat Race finishing post

bridge at the time. The superstructure is faced with Portland stone, apart from the lower portions of the arches, which have been hammered to give the concrete faces a pleasing appearance. The Thames Conservancy specified that the bridge should allow 25 feet of headroom above Trinity High Water (THW). THW was defined as the mean level of water at the highest spring tide (today, the technical name is Mean High Water Spring). In order to achieve this headroom without any steep inclines, it was necessary to raise the approach road some way back from the river. Therefore on each shore there is a further arch to provide for the riverside roads to pass underneath. These shore spans are divided up with columns between the carriageway and footway in an aesthetic design, and are flanked by flights of steps on either side from the high level of the bridge to the river level. Chiswick Bridge is certainly an elegant structure and Sir Herbert Baker's architectural contribution was much praised at the time.

Sir Herbert Baker (1862–1946)

Herbert Baker was born in Cobham, Kent, of a large farming family. His skill at drawing led him to train as an architect. During his early career in London, he met Edwin Lutyens and they went on sketching trips together. In 1892, he moved to South Africa, where he built many private houses and public buildings, including the main government offices in Pretoria for the South African President Jan Smuts. Here he developed the 'grand manner' style of architecture typical of the late British Empire. In 1912, he went to India to work with his old friend, Lutyens, on the massive government complex at New Delhi. Here they succeeded in transposing English classicism to a tropical climate by including elements of local Indian architecture in their designs. After the First World War, Baker returned to London. He designed India House at Aldwych, South Africa House in Trafalgar Square, and Church House, near Westminster Abbey. His final great commission was to rebuild the Bank of England, where he constructed a massive seven-storey structure on top of Sir John Soane's perimeter wall. His long and distinguished career resulted in his being knighted in 1926.

The fact that Chiswick Bridge was designed by an architect of distinction could be said to mark a radical change in the relative influence of architect and engineer on bridge construction. The nineteenth century had been the age of the great engineers, who took the lead in designing their iron and steel structures. *Country Life* of 8 July 1933 was enthusiastic about this change in favour of the architect:

> The engineer, by his training, is a functionalist. If his figures work out correctly, the result for him is *ipso facto* beautiful. The architect, on the other hand, is trained to

regard his work in relation to its setting, and so to take a broader view than that which sees a structure in isolation as the solution of a problem.

The magazine praised the elegance of Chiswick Bridge as 'reflecting in its general design the eighteenth century Palladian tradition of Lord Burlington's famous villa at Chiswick'. Not everyone will agree that all the bridges designed by engineers are inappropriate for their setting any more than that architects' designs are always sympathetic to their environment. In any case, the remark about Lord Burlington's villa seems strange, as his villa is some way distant and completely hidden from the river.

Barnes Railway Bridge

The present Barnes Railway Bridge has been described as the ugliest bridge on the River Thames. It is known today mainly for being the bridge under which the Oxford and Cambridge rowing boats shoot towards the end of the annual Boat Race. Barnes itself even has one light-blue oar for Cambridge and one dark-blue oar for Oxford on its coat of arms. By the time the boats have reached the bridge, the winning crew is usually clear, although there have been a few close races, including one tie. The race dates back to 1829 when it was first rowed at Henley. The Putney-to-Mortlake course has been used annually almost without a break from 1845. The finishing post is just downstream of Chiswick Bridge.

The bridge crosses over the river on the south side from the middle of Barnes Terrace. This is still an attractive stretch of the river, with its fine Georgian buildings. It has been the home of many notable Barnes residents, including the composer Gustav Holst. Further upstream, by the Boat Race finishing post, is Mortlake, known in former times for its tapestry works, which were set up by Sir Francis Crane in 1619 under the patronage of James I. The aim of the works was to rival the weavers of Brussels, who had been the pre-eminent designers and makers of tapestry since the sixteenth century. With the slump in demand for tapestry, the factory closed in 1703 and the area went into decline. By 1811, there were only 180 houses in Barnes. The village had not grown in the eighteenth century nearly as much as its royal neighbours, Kew and Richmond. Therefore there was not sufficient population or demand to allow for the building of a bridge until the railway age.

The railway first came to Barnes in 1846. This was the London and South Western Railway from Nine Elms to Richmond, which was in 1848 extended to Windsor across Richmond Railway Bridge. The original scheme also included a 'loop line' from Barnes across the Thames to Chiswick and Hounslow to rejoin the Richmond-to-Windsor line at Feltham. Joseph Locke was appointed engineer for Barnes Railway Bridge, as he had been for the bridge at Richmond.

Locke's first problem was how to embank the river so as to provide a clearance of 21 feet at Trinity High Water without too much disruption to the elegant setting of Barnes Terrace. He managed to avoid the demolition of any important people's houses on the Terrace, although some property must have been affected. The *Barnes and Mortlake Herald* of 9 February 1935 records Abraham Badham's story of how his family lived in a small house here which had to be demolished for the construction of the railway. Demolition was delayed for a month until Abraham was born and mother and baby were in a position to move home. It seems that not all Victorian entrepreneurs were completely heartless.

Like the railway bridge at Richmond, Barnes Railway Bridge was an attractive structure of three cast-iron spans. The spans were each 120 feet long, as opposed to the Richmond spans which were 100 feet. This gives an idea of how the river grows wider as it flows eastwards. Fox, Henderson & Co. were the contractors, and they completed the project in time for it to be opened on 22 August 1849. As was frequently the case with railway bridges, there was no opening ceremony.

Not everyone agreed about the attractions of the new bridge. Barnes Terrace was one of the most fashionable parts of Barnes, and the locals living there were none too happy that a railway should pass through their midst. The proprietor of the White

Hart, Mr Will Winch, whose hotel was at the upstream side of the bridge, wrote in his account of Barnes, 'While you sit there, gentle reader, enjoying the quaint old view of Mortlake riverside, and perhaps anathematizing the railway bridge which blocks the view downstream . . .'[8]

By the end of the nineteenth century, engineers were becoming aware of the long-term instability of cast-iron structures. In May 1891, a cast-iron span of the Brighton line bridge at Norwood Junction collapsed, although driver skill avoided a serious accident. Therefore, in July 1891, Royal Assent was obtained for an Act allowing the LSWR to rebuild Barnes Railway Bridge with wrought iron. The contractors, Head, Wrightson & Co., started work in 1894 to extend the brick abutments and piers on the downstream side and then to rebuild the bridge with wrought-iron bowstring girders. An 8-foot-wide footpath was added to allow pedestrians to cross. This also provided room for spectators of the Boat Race. The LSWR saw an opportunity for profit and sold tickets for 15 shillings to view the end of the race from this vantage point. Sadly, today the

Barnes Railway Bridge with its bowstring arches

71

footbridge is closed for safety reasons during the race. To allow services to continue during the rebuilding, the spans of Locke's bridge were left on the upstream side. They still remain today, although they have not been used since then. They somewhat relieve the ugliness of the new bowstring girders. The new bridge was opened on 6 June 1895.

On 2 December 1955, a tragic accident on the bridge saw 13 people killed. A late-evening passenger train to Windsor ran into a goods train coming the other way. The cause was identified as a mistake by the Barnes signalman, and this led to the old semaphore signals being replaced by coloured lights controlled from a new signal box. Unfortunately, this measure has not provided complete safety, as later railway accidents have shown.

CHAPTER 4

Hammersmith

Together with Tower Bridge, Sir Joseph Bazalgette's Hammersmith Bridge is the most ornate monument to Victorian engineering on the River Thames. It crosses the river on one of its sharpest bends some three and a half miles west of Hyde Park Corner. Millions of people will know this idiosyncratic structure from television broadcasts of the Oxford and Cambridge Boat Race, when the cameras pick out the crowds by the bridge watching the two crews straining to pass underneath first. Although Hammersmith Bridge is less than halfway to the finish at Mortlake, the race is often won or lost by this stage. Only if the trailing boat is on the Surrey side of the river does it have much chance of catching up, as it will shortly have the advantage of being on the inside of the Surrey bend.

Not everyone has agreed on the aesthetic merits of this bridge. William Morris, who owned a riverside house in Hammersmith, called it simply 'this ugly suspension bridge'. However, it is now such a historic landmark that any attempt to remove it would cause widespread protest. Built in 1887, the present bridge replaced an earlier one on the same site designed by William Tierney Clark in 1827. At the time, Tierney Clark's bridge had aroused similar mixed emotions.

Hammersmith, which lies on the north of the river at this

point, is today part of the London Borough of Hammersmith and Fulham. The name is first mentioned in 1294 as 'Hammersmyth', being a combination of 'hammer' and 'smithy', presumably indicating the presence of blacksmiths here. However, by the eighteenth century, the area was covered by nurseries and market gardens which supplied fruit and vegetables to the growing population of London. The most famous of these was the Vineyard Nursery, which cultivated exotic plants, such as the newly imported fuchsia, and employed 200 people by 1824. The riverside location of Hammersmith so near to London attracted aristocrats and wealthy merchants, and many of their Georgian houses still line Upper Mall and Dove's Passage, upstream of Hammersmith Bridge. During the eighteenth century, road transport from London was much improved with the introduction of turnpikes. It was at this time that the road from London's Hyde Park to Hammersmith and on to the west of England became known as the Great West Road because of the considerable traffic passing along its route. As a result, the population of Hammersmith grew to 5,600 by 1801 and to over 10,000 by 1831.

A most remarkable set of events in 1820 brought the previously quiet suburb to national prominence. George IV had just succeeded his father, George III, as King. He had been separated from his wife, Caroline of Brunswick, who had lived abroad for several years with her Italian servant, Bergami. On hearing of the death of George III, Caroline returned to England to claim her position as Queen. Rejected by her husband, she came to live in Hammersmith in Brandenburgh House, which was sited just downstream of today's bridge. A Bill was introduced in the House of Lords to dissolve the royal marriage on the grounds of Caroline's adultery with Bergami, but it failed. Popular feeling was so much on her side that crowds flocked from all over London to Brandenburgh House to witness the many 'congratulatory addresses' delivered in her support. Despite all this adulation, which was probably more due to the general feeling that George IV had treated her badly than to any merit of her own, Caroline was turned away from the coronation ceremony at Westminster Abbey the following year. She died shortly afterwards. Thomas Faulkner, Hammersmith's first true historian, described the situation of Brandenburgh House in 1839: 'South of Hammersmith Bridge on the Middlesex side is the site of Brandenburgh House, once the seat of gaiety and fashion . . . It is now derelict and no longer attracts visitors' attention.'[9] Brandenburgh House was then pulled down. From an architectural point of view, this was a pity, as the juxtaposition of this Jacobean mansion with the Tuscan towers of the soon-to-be-built suspension bridge would have created a sensational view. A distillery was later built here and remained until the 1960s. Today the Riverside Studios occupy part of the site.

It was inevitable that demand for a river crossing to the Surrey side would at some stage prove overwhelming. As far back as 1724, Daniel Defoe had written that Hammersmith aimed to obtain the grant of a market and that there was:

> Some talk also of building a fine stone bridge over the Thames; but these things are yet but in embryo, tho' it is not unlikely but they may both be accomplished in time, and also Hammersmith and Chiswick joining thus, would in time be a city indeed.[10]

By the early nineteenth century, with the increase in population and improvements in transport from London, the situation was much more favourable to the project of constructing a bridge. People who wanted to cross the river from here to the Surrey side had to make a five-mile detour via either Kew or Putney Bridge. Pedestrians had the choice of walking to Chiswick Wharf, where there was a ferry crossing to Ferry Lane in Barnes, or of hiring one of the watermen who frequented the Black Lion pub near the West Middlesex Waterworks.

Although it was rejected, the initial idea for building a bridge at Hammersmith came from the eccentric engineer Ralph Dodd. Finance for Dodd's proposal was to be raised by tontine, just as at Richmond. Evidently there was strong local support, but an insurmountable problem arose when Henry Hugh Hoare of Hoare's Bank refused to sell the strip of land needed to construct the approach road on the Surrey side. The proposal had to be abandoned, much to the gratification of the proprietors of Kew and Putney bridges. The latter had strongly opposed the idea because it would have introduced unwelcome competition and reduced their own profits.

Ralph Dodd (c.1756–1822)

Dodd's entry in the *Dictionary of National Biography* must be one of the saddest ever written. It starts with a question mark about the real date and place of his birth, which is stated to be either South or North Shields. Almost everything he turned his hand to ended in failure. His first major engineering venture was to promote a tunnel under the Thames from Gravesend to Tilbury. This ambitious project lasted from 1799 to 1802, when it had to be abandoned. His scheme for the Grand Surrey Canal was no more successful. The plan was to link the market gardens of Surrey to London by a canal from Epsom to the Greenland Dock at Rotherhithe. The construction project never advanced much further than building a long extension dock leading out of the Greenland Dock. Dodd produced design proposals for many other major construction projects which were eventually implemented, but not by him. These included designs for the new Waterloo and the replacement London bridges, both of which were taken over and completed by John Rennie. Dodd did finally manage to build an iron bridge over the River Chelmer at Springfield in 1820. If only half his other schemes had come to fruition, he would stand with Telford and Brunel as one of the great British engineers of the Industrial Revolution.

However, the proprietors of Kew and Putney bridges were not to enjoy their triumph for long. A group of local people formed the Hammersmith Bridge Company and raised £80,000 with a view to presenting another Bill before Parliament. The difference this time was that agreement was reached with Mr Hoare that the company would purchase his whole estate rather than just the land needed for the Surrey approach road. Strong opposition was again voiced in parliamentary debates, mainly inspired by the proprietors of Kew and Putney bridges. When it was clear that the Bill would be passed, they attempted to obtain compensation for the damaging effect the new bridge would have on their businesses. This was rejected largely because it was felt that they had already made exorbitant profits out of their virtual monopolies. As mentioned in Chapter 2, the impending construction of Hammersmith Bridge and its likely impact on the profits from Kew Bridge was probably the reason why the proprietor, Robert Tunstall, decided to sell his stake in Kew Bridge soon afterwards.

The Act enabling the building of Hammersmith Bridge, which was to be the first suspension bridge over the River Thames, finally received Royal Assent on 9 June 1824. The Act incorporated the Hammersmith Bridge Company and appointed the first management committee members. It also laid down a complicated set of rules covering the duties of the committee and shareholders' voting rights. Shareholders who were infants had a vote through their guardians, and 'lunatics' through their 'lunatic committee'. Females could vote only through a male proxy. Tolls were also laid down. These must have confirmed the worst fears of the proprietors of the other bridges, as they were cheaper for every category except pedestrians. A local newspaper published a table of comparison from which this is an extract:

	Hammersmith	*Putney*	*Kew*
Foot passenger	½ d.	½ d.	½ d.
Ass unladen	1 d.	2 d.	2 d.
Horse	1½ d.	2 d.	2 d.
Horse and chaise	4 d.	6 d.	8 d.
Carriage and two horses	6 d.	1 s.	1 s.

Punishments were set for any damage caused to the bridge. The standard fine was five pounds. In addition, those convicted of damage had to pay for the necessary repairs. This seems reasonable compared with the severity of the punishments laid down in the Richmond Bridge Act of 1774, which provided for deportation to America. One of the most important provisions of the Act covered the approach roads. Improved communications between London and the west was one of the main justifications for building the bridge. On the Middlesex side, the approach road was scheduled to connect directly to the Great West Road. On the Surrey side, a much longer road was required to pass through Mr Hoare's estate to Barnes Common and then on to the Upper Richmond Road.

The engineer chosen to design Hammersmith Bridge was William Tierney Clark, who had worked closely with the

William Tierney Clark
plaque on the site of the
Middlesex Waterworks

Hammersmith Bridge Company on the proposal. Like Ralph Dodd, the original proponent of a bridge here, Clark was involved with the West Middlesex Waterworks. In fact, he had been appointed chief engineer. Clark's proposed design for a suspension bridge at Hammersmith was attractive as it required the construction of only two river-piers and provided a 400-foot-wide navigation path for shipping.

William Tierney Clark (1783–1852)

William Tierney Clark was born in Bristol, where he served his apprenticeship to a millwright. He then moved to Coalbrookdale, where he was employed in the ironworks of Abraham Darby, constructor of the world's first iron bridge, over the Severn Gorge in 1779. In 1808, John Rennie visited the ironworks and was so impressed by the young Tierney Clark that he offered him a job at his own works at Blackfriars. There he supervised several of Rennie's projects in cast-iron construction. When the job of engineer to the West Middlesex Waterworks came up, Rennie was happy to recommend him. The company was established in 1806 and many of the personalities involved were also connected with the Hammersmith Bridge project. In 1818, the waterworks company allowed him to practise as a consulting engineer, and his first major project was Hammersmith Bridge. In 1829, he took over the project to construct another suspension bridge on the River Thames, at Marlow. Marlow Suspension Bridge lasted until after the Second World War, when it was reconstructed so as to retain the appearance of Tierney Clark's original bridge. Fortunately, this allows us to see what the first Hammersmith Bridge looked like, albeit on a smaller scale.

Tierney Clark's *magnum opus* was undoubtedly the famous chain bridge over the Danube at Budapest. This was also a suspension bridge, and the first bridge to span the fast-flowing 1,500-foot-wide river between the two previously separate towns of Buda and Pest. The bridge was completed in 1849 and survived until its destruction by the retreating German army in 1945 at the end of the Second World War. After the end of the war, the bridge was rebuilt according to Tierney Clark's original design, and it stands today as a foreign monument to the great engineer. Tierney Clark died in 1852 and was buried in St Paul's Church, Hammersmith. His memorial, engraved with the image of a suspension bridge, can be seen there today.

The choice of a suspension bridge was a daring decision to take, since no successful large-scale suspension bridge had ever been built except for the pioneering Union Bridge over the Tweed near Berwick, constructed in 1820 by Captain Samuel Brown (1776–1852). Previous suspension bridges had not been stiff enough to keep the platform stable in high winds or when there was heavy traffic or people movement. Brown had invented a system using solid, straight eye-bars joined together in order to form a firm overhead structure from which to hang the rods which hold up the road platform. Brown supplied the ironwork for Hammersmith Bridge, but it was Tierney Clark who designed it with two massive stone river-towers which supported the suspension chains and formed a Tuscan archway through which the road platform ran. There were no towers on the banks of the river. The chains disappeared under the octagonal toll-gates located on each end of the bridge and were anchored firmly in lengthy abutments underneath the road.

Since Thomas Telford (1757–1834) was in the process of constructing a similar suspension bridge over the Menai Straits between Wales and Anglesey at this time, Clark submitted his plans for Telford's comments. There was considerable mutual respect as well as rivalry between the great engineers of the nineteenth century, and so it was not surprising when they asked each other's advice. In this case, Telford was paid a consultancy fee of £52 10 s. Further cooperation between the two resulted from Telford's work on his only London project, at St Katherine's Docks. Some of the spoil excavated to form the dock basins was shipped to Hammersmith to fill in the marshy land on the Surrey side of the river so as to support the approach road. The rest of the excavated spoil was used by the entrepreneur Thomas Cubitt to fill in the area of marshy land between Buckingham Palace and Chelsea before developing it for the landowner, Lord Grosvenor. The resulting development, known as Belgravia, is still part of the Grosvenor estate and is one of London's most fashionable residential districts.

Telford's Menai Bridge was completed in 1826, one year earlier than Hammersmith Bridge. The Menai Bridge has a central span between the supporting towers of 579 feet. However, the road between the towers and the shore is supported on masonry arches. At Hammersmith, the central span between the river-towers is 400 feet but the suspension chains also support the road platforms between the river-towers and the riverbank. This gives a total length of 688 feet and allows the claim that Clark's Hammersmith Bridge was the

Hammersmith Bridge of 1827

longest suspension bridge in the world at the time it was built.

During construction, crowds flocked to watch as the pioneering bridge emerged from the river. One of the most assiduous visitors was George IV's younger brother Augustus Frederick, Duke of Sussex, who often walked there from his 'Smoking Box', which today is called Sussex House, situated nearby on the riverside in Dove's Passage. It was he who laid the foundation stone on 7 May 1825. According to the invitation ticket, this was done in a Masonic ceremony. Cofferdams had already been installed for the construction of the pier foundations and an amphitheatre was constructed inside so that the foundation stone could be suspended in readiness for the Duke to lower it into the cavity together with a bottle containing coins of the realm and a brass plate commemorating the event. The Duke was also invited to conduct the opening ceremony, which took place on 6 October 1827. For some reason, much debated by local historians, he declined. However, this does not seem to have dampened the enthusiasm of the crowds, who were greeted by the firing of cannons and treated to a firework display as the ceremony ended.

The new bridge attracted sightseers from all over London and this increased the takings from the tolls. It was highly praised in the press. According to the *Franklin Journal and American Mechanics Magazine* of April 1828, 'The architectural beauty of the masonry is a great improvement to the hitherto clumsy masses of stone in the other erections of a similar description, and the whole edifice forms a highly ornamental feature to the River Thames.'

From a practical point of view, the bridge had significant shortcomings. The width of the carriageway was 20 feet, and there were two footpaths of 5 feet on either side. This was not unreasonable for the traffic conditions at the time, except that where the road went under the thick stone arches, its width was reduced to only 14 feet and at this point had to provide for both vehicles and pedestrians. Traffic was about to increase substantially, not least because of the existence of the bridge itself. It could even be said that the bridge put Hammersmith on the map rather than vice versa. Different modes of transport were also about to come on-stream. The year 1829 saw not only the first Oxford and Cambridge Boat Race but also the invention of the horse-drawn omnibus. In that year, George Shillibeer instituted an omnibus service from Paddington to Liverpool Street, and soon omnibuses were bringing passengers to Hammersmith and beyond. There was certainly not enough room for two-lane traffic if an omnibus was passing under the arch of one of the river-piers. A further problem arose from the fact that the roadway was only 18 feet above THW, which was much too low for steamers, with their tall funnels, to pass. Fortunately, George Dodd, the son of Ralph Dodd, had invented a method of lowering the funnels, but this could cause some discomfort to the passengers and in 1842 was lampooned in *Punch*:

> A vessel passing under the bridge is compelled to lower its chimney onto the heads or into the laps of the passengers, besides rendering it incumbent on all on board to bend to circumstances by placing their heads between their knees during the time occupied in passing under the elegant commodious structure.

By 1843, the Hammersmith Bridge Company decided that there was money to be made out of the steamboat traffic, and they installed a steamboat pier against the Surrey-side river-tower. This is considered to be the first structural feature on a bridge to recognise the coming of the steam age.

A major change in London's government came in 1855 with the passing of the Metropolis Management Act. This set up the Metropolitan Board of Works, with responsibility for central services, including extensive powers over the management of bridges. Local boards were set up to run local affairs, and Hammersmith came under the Fulham District Board of Works. The scene was now set to bring the Thames bridges under public control. Already in 1840 a petition had been submitted by the Metropolitan Anti-Bridge Toll Association to free Waterloo and

Vauxhall bridges from tolls. The petition concentrated on Waterloo Bridge, which people avoided because of the tolls, causing crowding on the free bridges which were under the control of the Corporation of London. This caused delays and made people late for work, as they would walk miles to avoid the tolls. With the creation of the MBW, pressure grew to free all the bridges in its area of control, especially since the upstream bridges from Kew to Staines had already been freed, as described in Chapter 2. In 1877, the Metropolis Toll Bridges Act was passed to allow for the MBW to purchase the bridges and abolish the tolls. Hammersmith Bridge was eventually sold to the MBW in 1880 for £112,000. Hammersmith, Fulham and Wandsworth bridges were all declared toll-free on the same day, 26 June 1880. The Prince of Wales attended the celebration ceremonies, which drew wildly enthusiastic crowds.

By now, Hammersmith Bridge was causing increasing concern, because of both its narrow width and the heavier loads of traffic. With the extension of the Metropolitan Line to Hammersmith, the population received a further boost on the Middlesex side of the river. On the Surrey side, where there had been mainly open ground when the land was bought from Henry Hoare in 1824, major development had taken place along the road to Barnes Common.

The general increase in traffic was exacerbated on the day of the Boat Race, when, by the 1870s, over 11,000 people crammed onto the bridge and many climbed on the chains to get a view. A famous picture of the crowding on Hammersmith Bridge at the 1862 Boat Race, which was won by Oxford, was painted by 16-year-old Walter Greaves, the son of a local boat builder. Sadly, he never made a living out of his art, and died in poverty in a charity home. Today, this painting hangs in Tate Britain and has spawned thousands of postcards and reproductions. A letter from a member of a Cambridge crew who had rowed in the Boat Race during the 1870s appeared in *The Times* of 17 March 1922, describing the atmosphere as the boats passed under Hammersmith Bridge: 'Of the bridge itself little or nothing could be seen, but only a mass of humanity, which, like a swarm of bees, hung itself on every available point, high or low, not counting risk to life or limb.' By 1876, Hammersmith Bridge was closed on the day of the Boat Race because of fears for its safety as well as the safety of the crowds. Later that year, a strange rumour spread through London that the bridge had collapsed and that many people had drowned. The rumour was given credence because of the earlier closure, and by 5 p.m. thousands had travelled to Hammersmith to see the non-existent spectacle.

Reports on the safety of the bridge were produced by Rowland Mason Ordish, the designer of the Albert Bridge, and by John Hawkshaw, the designer of Hungerford Bridge, in 1869 and 1878 respectively. Both found some faults but in general were impressed by its strength and declared it in good enough shape to carry current levels of traffic. However, in 1882, a boat collided with the bridge and an investigating policeman fell through a hole in the walkway into the river. Sir Joseph

1866 scene of the Boat Race crowds on the old Hammersmith Bridge

Bazalgette, the chief engineer of the MBW, which by this stage owned the bridge, had already raised safety concerns. He now produced a report recommending the complete reconstruction of the bridge superstructure on top of the existing pier foundations, after the Surrey-side pier had been underpinned. The shareholders of the Hammersmith Bridge Company must have been relieved to have sold out just two years before. An Act was passed in August 1883 enabling the reconstruction of the bridge along the lines recommended by Bazalgette, as well as the building of a temporary wooden bridge to cope with cross-river traffic until the work was complete. Needless to say, Bazalgette himself was chosen to design the new bridge.

The new Hammersmith Bridge, like the old, was designed on the suspension principle, but it has a much more fanciful appearance than its predecessor. Structurally, there are major differences in the use of materials. The suspension chains are of

Bronze bust of Joseph Bazalgette, located on the Embankment near Hungerford Bridge

steel rather than wrought iron. The river-towers, instead of being built of stone, have frames of wrought iron which are clad in ornamental cast iron. Since iron is lighter than the equivalent strength masonry, the towers take up less space and allow a wider opening for road traffic through the arches. As a result, the carriageway under the arches is now 21 feet wide, instead of 14 feet, and there is room for two 6-foot-wide footways, which are cantilevered and curl round the outside of the towers, rather than sharing the carriageway as was the case with the old bridge. On the riverbanks, instead of the toll-gates which had been located there when the old bridge was built, Bazalgette constructed highly decorative abutments which take the suspension chains underground to a depth of 40 feet, where they are firmly anchored. The most striking of the many ornamental cast-iron features are the coats of arms to be found on the sides of the abutments. The MBW was not granted an official coat of arms but used an unofficial design with the Royal Arms in the centre surrounded by the arms of the various civic authorities that came under their auspices. Unfortunately, Hammersmith did not receive its coat of arms until 1897, and so it is not represented on its own bridge.

The contractors were Messrs Dixon, Appleby and Thorne, who completed the work in less than three years at a cost of £82,177. The opening ceremony on 18 June 1887 seems to have been a much less splendid affair than the ceremony for the freeing of the tolls. Hammersmith Vestry refused to put up any street decorations, as it considered that this would be a waste of public money. Instead of the Prince of Wales himself, his son, Prince Albert Victor, conducted the ceremony before going on to open Battersea Bridge, also designed by Bazalgette. One further addition to the new bridge occurred in 1894 when a new steamboat pier was constructed. Shortly after this, steamboat services started to decline and the pier was removed in 1921.

The new Hammersmith Bridge has certainly had an eventful life over the 130 years of its existence. Its owner, the MBW, was replaced in 1889 by the London County Council (LCC), which was responsible for it until 1964, when the LCC was abolished

Ornate cast-iron coat of
arms at the south end of
Hammersmith Bridge

and the newly created Greater London Council (GLC) took
over. Finally, in 1986, the Prime Minister, Margaret Thatcher,
abolished the GLC, and the bridge returned to local control,
that of the London Borough of Hammersmith and Fulham,
which still manages it today. Politics were involved in all these
changes, and the Labour-controlled borough council could not
help making capital out of the fact that the Conservative
government had abolished the GLC one year before
Hammersmith Bridge's centenary, by promising to celebrate the
occasion in style itself. The most significant preparation for the
ceremony was the repainting of the bridge. This proved a
challenge, as no fewer than 46 coats of paint were found
underneath the current battleship-grey colour, some of them
dating right back to when the bridge was built 100 years before.
In fact, the bridge has changed colour a number of times over

its lifetime since the initial green-and-gold scheme designed by Bazalgette. The new colour scheme for the centenary was designed by Ken Mellors, the GLC architect, and approved as authentically Victorian by the Royal Society of Fine Arts. Unfortunately, it was not possible to paint over the existing paintwork, as rust had permeated into the ironwork in places, so all 46 coats had to be blasted off down to bare metal and 7 new coats applied. The result was certainly very attractive, but unfortunately it was considered too expensive to use gold leaf for the gilding, and so gold paint was used instead. Bazalgette himself had specified gold leaf for the bridge, but nowadays this is seldom used for large-scale exterior decoration. One place where it is possible to see true gold-leaf gilding is on the insignia on top of Sea Containers House near Blackfriars Bridge.

Despite the criticisms of William Morris and others, the bridge has proved to be a much-loved structure. Unfortunately, it has suffered from problems similar to those that beset Tierney Clark's bridge. As the river-towers were constructed on the same foundations, it is still really far too narrow for modern traffic conditions. Moreover, it has had to be strengthened a number of times. In 1973, major reconstruction was required, including replacement of the timber decking under the road surfaces and repairs to and strengthening of much of the ironwork. As a result, the original lamp-posts which straddled the wrought-iron girders had to be replaced with fibreglass replicas, which do still look authentic. Further major repairs were initiated in 1997, and the bridge was closed while these were carried out. Indeed, there was even doubt as to whether Hammersmith Bridge would ever reopen for traffic apart from buses. A local referendum was held, as many local people were pleased at the reduction in traffic where they lived. Cyclists and bus companies were definitely opposed to the reopening, as they had exclusive use of the carriageway when the repairs were completed in 1999. Businesses, however, were strongly in favour, as fewer customers had been coming into the area while the bridge was closed. In the end, people voted three to two in favour of reopening the bridge to general traffic. Weight restrictions were imposed on the reopening, as indeed had been the case since the first concerns were raised in the 1960s. Since the vote in favour had been less than overwhelming, the bridge was reopened without any ceremony.

Natural deterioration has not been the only danger to which Hammersmith Bridge has been exposed. The IRA has tried to blow it up on no fewer than three occasions. The first was on 29 March 1939, when, in the early morning, Maurice Childs was walking across the bridge and saw smoke and sparks coming out of a suitcase. Bravely, he opened the suitcase and saw that there was a bomb inside. Immediately, he picked it up and threw it into the river. Shortly after, it exploded and threw up a 60-foot column of water. A second device did explode and caused some damage to the suspension chains and rods, as well as to nearby houses. For his brave act, Childs was awarded an MBE.

The events surrounding the terrorist attack were most

Hammersmith Bridge today

extraordinary. Two members of the IRA had hired a car and forced the chauffeur, Mr Moffat, to drive them to the bridge. Once the bombs were in place, they made Moffat drive them to Putney Bridge, where they left him and walked on across the bridge, presumably thinking he would not dare follow them. As it happened, a policeman came by, and he and Moffat gave chase. The men were caught and Moffat had no difficulty in identifying them in court. They were sentenced to a total of 30 years' penal servitude between them. This could be considered the first action of the Second World War, as the war officially started a few months later and it is known that the IRA in general supported Germany because of its hatred of Britain. The enemy bombers were far less effective in damaging Hammersmith Bridge than was the IRA.

The second IRA attempt occurred in 1996 when two of the largest Semtex devices ever planted in mainland Britain failed to go off. In June 2000, the IRA tried again, and this time a hole was blasted in one of the girders at the south end of the bridge. The repairs were supposed to be completed by September 2000, but completion was delayed until December because of a catalogue of errors. The wrong sort of steel had been delivered from Belgium and was returned for replacement, and problems with the newly relaid road surface meant that the work had to be redone. This caused the *Evening Standard* of 18 September 2000 to publish a banner headline, 'A Bridge Too Far and It's Traffic Mayhem'. The article listed a number of problems on other bridges as well as the delays on Hammersmith Bridge. Kew Bridge, Kingston Bridge and Westminster Bridge were undergoing repairs at the same time, and Putney Bridge had introduced a 24-hour bus lane. Following the eventual completion of the repairs, 20 CCTV cameras were installed with a view to preventing any further terrorist attacks.

Hammersmith Bridge will always be remembered for a singular act of heroism. As reported in the *West London Observer* of 2 January 1920, Lieutenant Charles Campbell Wood had been walking with a friend, Mr Monk, across Hammersmith Bridge late at night on 27 December 1919. They both heard a scream and saw a woman fall from the parapet into the river. Lieutenant Wood dived straight in after her and brought her safely to the riverbank. The woman had had an argument with her daughter and apparently tried to commit suicide. She survived. Wood had dived from 30 feet into icy water only 12 feet deep and had injured his head. The doctor who examined him in hospital said that he could not understand how he was able to retain his senses and swim ashore with the woman, as he collapsed immediately afterwards with serious head injuries and in a state of shock. He died soon after from his injuries. His friend said that he had plunged into the water without any thought of danger to himself. Wood was born in South Africa, had joined the Boy Scouts and St John Ambulance, and was an accomplished sportsman. He had come to England in 1917 to serve in the RAF as a pilot. As the coroner stated, a man who lived life to the full had sacrificed himself for someone who apparently did not wish to live. He received the posthumous award of a bronze medal from the Royal Humane Society and is buried in St Marylebone cemetery, where his tombstone still stands.

Despite its history and landmark status, Hammersmith Bridge has in the past been threatened with replacement by a more modern structure capable of handling multiple lanes of traffic. The 1926 Royal Commission on Cross-river Traffic commented on the frequent repairs to the bridge, which resulted in congestion as traffic had to be reduced to a single lane. The Commission recommended that the bridge should be pulled down and rebuilt to carry four lanes of traffic. No final decision was taken at the time, although the recommendation was revived on several occasions. No serious attempt to replace the bridge has been made since the 1960s. However, when the bridge was closed for refurbishment in 1996, Brian Sewell wrote an article in the *Evening Standard* of 11 February resurrecting this idea. He described the bridge as a 'monument to low technology rooted in the age of Isambard Kingdom Brunel devised a decade before the motor car that has now . . . brought it to its knees'. He went on: '. . . as it is a much-loved structure for its whimsical charm, it should be redone as a footbridge elsewhere over the Boat Race course using lottery money. This would allow a new bridge fit for modern traffic to be built in its place.' Today, no one would seriously contemplate the removal of Bazalgette's bridge, as it would probably lead to revolution on the peaceful riverside at Hammersmith.

CHAPTER 5

Putney and Wandsworth

Putney Bridge crosses the Thames between Fulham on the north bank and Putney on the south bank. Today's bridge replaced the old wooden bridge known as Fulham Bridge, which was opened in 1729. This had been the first bridge constructed across the tidal Thames in London since the thirteenth-century Old London Bridge. At the time, it was the only bridge between London Bridge and Kingston, where a bridge had existed from medieval times. Wandsworth, on the other hand, had no bridge until 1873. Wandsworth Bridge was then built one mile downstream from Putney Bridge. It lasted only 65 years and was then replaced by the present steel structure in 1938. In between them is Putney Railway Bridge, built in 1889 to take the District Line over the river to Wimbledon.

Putney Bridge
The settlement of Fulham has a long history. In the eighth century, the Bishop of London, Waldhere, was granted the manor of Fulham together with the duty to maintain four bridges over two of the rivers that used to flow into the Thames in the area. The rivers, with their bridges, disappeared in the nineteenth century because of pollution, but the manor house still stands in its large landscaped grounds to the west of Putney

Bridge. Parts of the building date back to the sixteenth century, when Bishop Fitzjames built the Tudor courtyard. Later, the manor house became known as Fulham Palace, and it served as the summer residence of the bishops of London until they handed it over to the local council in the early twentieth century.

Fulham's importance was greatly enhanced by the presence of the bishops of London, many of whom went on to become archbishops of Canterbury and who frequently played a part in events which affected the history of the nation. Bishop Grindal once stood up to Queen Elizabeth I and told her: 'Although you are a mighty prince, the Lord in Heaven is mightier.' Perhaps surprisingly, he did not lose his head or even his position for setting up the Church against the monarch. Others were not so fortunate. Bishop Ridley was burned at the stake in Oxford by Elizabeth's half-sister, Mary, for refusing to acknowledge the supremacy of the Roman Catholic Church, and Archbishop Laud was executed on the orders of Parliament for opposing Puritan dogma.

Fulham started as a small settlement around the Bishop's Palace. Its original name, Fulhanham, means 'foul' or 'muddy' hamlet and arises from the large areas of mud deposited on the low-lying riverbanks by the ebb and flow of the tide. Local employment was originally concerned with fishing and growing fruit and vegetable produce to supply the increasing demands of London. In 1670, the potter John Dwight invented the process of making salt-glazed stoneware and set up his famous Fulham Pottery just to the east of today's Putney Bridge. The pottery lasted until the 1940s. In the eighteenth century, Fulham, like Hammersmith, became a favourite location for the rich to build mansions and villas where they could spend summer weekends away from the city. Most of these mansions are long gone, but Hurlingham House and Fulham House survive.

The bishops of London used the river to travel from London to Fulham Palace in their private barges from the earliest times. In addition, they owned a ferry which they leased to local businessmen, who in turn sublet it to ferrymen to transport pedestrians and vehicles between Fulham and Putney. The first record of a ferry here dates back to 1210. Since Fulham was on the main south-western route out of London, the ferry was busy, and many inns and taverns grew up to cater for travellers who wished to refresh themselves on the journey while their coaches were loaded onto the ferry. The most famous of these was the Swan Inn, which was situated by the ferry, about 100 yards to the east of today's bridge. The Swan Inn was built in 1695 and lasted until 1871, when it was burnt down, probably by an arsonist.

The ferry increasingly created a bottleneck, as well as often causing risk to life and limb. In 1642, during the Civil War, a bridge was built at Fulham for the first time. Charles I and Prince Rupert had advanced on London with the Royalist army from their headquarters in Oxford. They won a skirmish at Brentford against a smaller Parliamentary force, but retreated to Kingston when the Earl of Essex went to meet them with his main army of 24,000 men at Turnham Green. Since the Royalists controlled Kingston Bridge, Essex decided to build a bridge across the river at Fulham so that he could attack Kingston from the south. The bridge was constructed with boats tied together and with defensive earthwork fortifications on both riverbanks. It was never used in anger because the Royalists retreated back to Oxford before Essex could cross the river to attack them. The earthworks on the south bank remained until 1845.

The first attempt at creating a permanent crossing at Putney was made in 1671, when a Bill for the construction of a wooden bridge was introduced to Parliament. The Bill met strong opposition, and records of the debate make for entertaining reading. Opponents claimed that a bridge at Putney would jeopardise the prosperity of London and even annihilate the city altogether. They argued it would stop the tide and destroy the Thames as a navigable river. It would prevent wherries passing at low tide and this would affect the interests of the watermen on whom the nation depended for providing experienced seafarers in time of war. A Mr Boscawen complained that Putney Bridge would be the thin end of the wedge and could result in further bridges at Westminster, Blackfriars, Somerset House and Guildhall. He added, to great hilarity from his supporters, that some of these might even be built of iron. He ended with a peroration in which he ironically threatened to bring in Bills for bridges at Chelsea, Hammersmith and Brentford. Mr Boscawen sat down to great acclaim and the Bill was thrown out by 67 votes to 54. He would surely have turned in his grave to find that bridges were built at each of the locations he had identified within the next 100 years.

Despite the opposition from vested interests, pressure continued to mount for a second bridge over the Thames at Westminster, Vauxhall or Putney. Matters came to a head in 1720, when Prime Minister Robert Walpole was returning from a visit to George I in Kingston to attend a debate in the House of Commons. He rode on horseback with his servant to Putney only to find the ferry was on the other side of the river. The ferrymen were drinking in the Swan Inn and took no notice of Walpole's shouts for them to take him across the river on vital national business. It was hinted that they were all Tories opposed to Walpole's Whig Party, but ferrymen were notoriously heavy drinkers and may well have been in no fit state to answer his calls. In any case, Walpole had to take a longer way round to Parliament, and this incident seems to have made up his mind to have a bridge built at Fulham.

In 1726, with Walpole's support, an Act was passed 'for Building a Bridge across the River Thames from the town of Fulham in the County of Middlesex to the town of Putney in the County of Surrey'. Commissioners were appointed to manage the project and maintain the bridge, but, to avoid the possibility of corruption, they were not allowed to invest money themselves. Since it proved impossible to raise the money from other sources, an amended Act was passed in 1728 which allowed more financial flexibility. Thirty subscribers, of whom one was Walpole himself, invested £1,000 each in shares which gave the right to receive income from the tolls in perpetuity. The shares also gave the right to vote in both Middlesex and Surrey elections. It was not long before the shares were divided into fractions and the fractions sold off to people who wanted extra votes. Since far fewer people had the vote in those days, this was an important privilege, only abolished in 1864. The Act also laid down toll charges, which were similar to those at Richmond, and the punishment of a death sentence for anyone convicted of damaging the bridge.

Several designs for Fulham Bridge were submitted, including one for a bridge of boats, which was possibly inspired by the 1642 crossing constructed by the Earl of Essex. The commissioners chose a design by Sir Jacob Ackworth, who had designed a number of bridges on the upper Thames, including the bridge at Kingston. However, according to Thomas Faulkner, this design was considerably modified by Mr Chiselden, who was surgeon to Chelsea Royal Hospital and one of the 30 subscribers.[11] Another subscriber, Thomas Phillips, was given the construction contract. He completed the 768-foot-long wooden structure with its 26 narrow openings in only 8 months. The central span was built wider than the rest so as to allow more room for boats to pass through. It measured 30 feet and was known as Walpole's Lock in honour of Robert Walpole, who had helped ensure the passage of the Act in face of continuing opposition from vested interests.

The compensation paid to those whose income was affected amounted to over £9,000, which was nearly as much as the actual cost of the bridge itself. The Bishop of London, who had owned the ferry, received only £23. However, he and his household were granted free use of the bridge. This was open to abuse by people who were not actually members of the Bishop's household. Thomas Crofton Croker, in his *A Walk from London to Fulham*, wrote that: 'People were very much astonished at hearing the exclamation "Bishop!" shouted out by the stentorian lungs of bricklayers, carpenters or others who may be going to the palace, that being the password for going over free.' Others to receive compensation included the ferrymen who had lost their jobs. It is also recorded that Walpole agreed to recommend them for employment at the Custom House. This seems a generous gesture from the Prime Minister, who had a reputation for arrogance and corruption, especially in view of his anger at the behaviour of the ferrymen

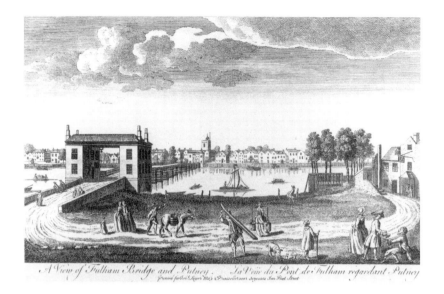

1760 engraving of Fulham Bridge showing the southern tollhouse

who had ignored his calls to take him across the river on a previous occasion.

Fulham Bridge was opened without ceremony on 29 November 1729. The first person to cross in a coach was the Prince of Wales, when going to and from Richmond Park for one of his regular hunting expeditions. The Prince had long been a supporter of the bridge and showed his delight to the workmen by donating five guineas. He could have had free passage across the bridge, since the King paid an annual £100 fee to the company to cover all crossings by members of his household. On the introduction of the new Gregorian calendar in 1752, when the Government abolished the dates between 3 and 14 September, the King deducted £1 10 s. from the £100 to compensate for the 11 lost days. However, the King was the loser in the following year. London bankers had always been liable to pay their taxes annually on 25 March, but they delayed their payment for 11 days until 5 April, which has marked the end of the tax year in Britain ever since.

Amazingly, the wooden bridge was to last for over 150 years, which was longer than many of the later stone bridges on the tidal Thames. Although the maintenance costs were high, the bridge proved a reasonable investment for the subscribers. By the time it was eventually sold, the income from the tolls had doubled, as had the value of the shares. Regarding the profitability of the bridge, Archibald Chasmore relates the following anecdote about Theodore Hook, a writer of farces, who owned a villa just upstream of the bridge:

> A friend was looking at the bridge from Hook's garden and said he had heard that the bridge was a good investment. 'I don't know,' said Theodore, 'but you have only to cross it and you are sure to be tolled.'[12]

The Fulham Bridge Company's profitability was threatened both by competition from other existing bridges and by proposals for new bridges. The original proprietors had promoted the benefits of their new bridge when their proposal had been attacked by the City Corporation, which had an interest in preserving the monopoly of London Bridge. But now that they owned a virtual monopoly themselves, they opposed every proposal for other new bridges upstream from Westminster. They failed to win these arguments, and they were no more successful in a dispute with the proprietors of Hammersmith Bridge over a directional sign they had put up outside the coach stand at Knightsbridge. This claimed that the shortest route to Richmond was via Fulham Bridge. The Hammersmith Bridge Company objected on the grounds that its own measurements showed that the route via Hammersmith Bridge was half a mile shorter. The Metropolis Road Office, which was responsible for road signs, came down on the side of Hammersmith, and the Fulham proprietors were forced to change their sign accordingly.

The profits from the bridge clearly depended on the amount of traffic, but also on the efficiency of the toll collectors. Tollhouses were located at both ends of the bridge and were normally manned by a resident manager, his assistant and three tollmen. From 1801, when Nathaniel Chasmore was appointed manager, until the freeing of the bridge from tolls in 1880, the managers all came from the same local family of builders. The last of them was Archibald Chasmore, who wrote *The Old Bridge*, which covers the history of Fulham Bridge until it was demolished in 1882. Before 1801, there were a variety of managers, not all of whom proved satisfactory to the proprietors. In 1739, the current manager, Robert Rawson, resigned after being hurt in a fight on the bridge. The proprietors refused to pay him compensation because they noted that income from the tolls increased substantially after he left. Since the manager was responsible for keeping accounts and checking the honesty of the tollmen, there was a great temptation to commit fraud, and the company may have suspected this. His successor seems to have devised an improved system of checks, and this resulted in increased takings. Unfortunately, the new manager in turn was later dismissed for using £86 of the company's money for his own purposes.

The tollmen worked day and night shifts seven days a week. Until 1835, they had no regular leave of absence. Absence due to sickness was treated relatively well for the time. It is recorded that in 1877 a Mr Robinson received half pay until he was fit to return to work. In 1761, a Mr Slane was killed by falling timber and his widow was allowed ten shillings a quarter towards her rent and two guineas towards her support. In fact, the secure wages and regular Christmas bonuses made the job comparatively attractive, despite its long hours. Tollmen did cause a variety of problems, and cases of absenteeism, dishonesty and drunkenness were not infrequent.

Problems also arose from rows between the tollmen and the public, not all of which were the tollmen's fault. In 1739, three young officers beat up the tollmen when crossing the bridge. A committee met to examine the tollmen's complaint. When the officers apologised and paid 20 guineas' compensation, it was decided not to prosecute them 'in consideration of their youth and the ruin that might follow in the course of prosecution against them'. During another incident, as recorded by the *London Evening Post* of December 1740, it was tollmen who were the aggressors, and two of them were indicted for assaulting two ladies and their servants. The prosecution alleged that the tollmen had tried to charge the ladies for more horses than they were taking over the bridge. The tollmen resorted to violence when the ladies refused to pay. They were found guilty, but their counsel 'moved for Mercy till next Term so that they could apologise and obtain pardon from the ladies'. The prosecution agreed as 'the aim was to make the Fellows an Example only for the Benefit of the Public'.

The eighteenth century was an unruly time and the tollmen were especially exposed to danger since the bridge led to Putney Heath and Wimbledon Common, both of which were notorious for highwaymen and robbers. The tollmen were armed with copper-headed staves, and bells were hung on top of the tollhouses so that they could warn each other of trouble. On one occasion, a tollman did benefit from criminal action when a coachman called Daniel Good threw him his coat as he passed through the toll-gate and asked him to keep it until his return. In fact, Good had just murdered his girlfriend in Putney; he was hanged for it and so never returned. Archibald Chasmore states that the tollman wore the coat until it was worn out.

Fulham Bridge was the scene of the attempted suicide of Mary Wollstonecraft in 1795. She suffered depression because of the neglect of her lover, the American writer Gilbert Imlay, and wrote him a suicide note stating her hope that she would not be rescued. She jumped from the bridge and lost consciousness when she hit the water but was resuscitated by some watermen who were passing at the time. After her rescue, she repeated that she had wished to die. She later married William Godwin, but died shortly after giving birth to a daughter, Mary, who married Percy Bysshe Shelley. It is thought that Mary Shelley was influenced in her creation of Frankenstein's monster by her mother's attempted suicide at Fulham Bridge, since the monster, like Mary Wollstonecraft, had life breathed into him against his will.

One of the most successful days for the tollmen was on 20 July 1867, when the Sultan of Turkey accompanied the Royal Family to a review of the Volunteers at Wimbledon Common. Carriages stretched from Hyde Park Corner over Fulham Bridge to the camp on Wimbledon Common in an unbroken line. Over £100 was collected and, according to Archibald Chasmore, not one base coin was taken in. Since counterfeiting was prevalent at the time, this was a considerable achievement.

Wimbledon Common reviews by royalty were popular, and very profitable for the proprietors of Fulham Bridge. On 11 June 1845, an advertisement appeared in *The Times,* announcing a grand review of the Life Guards on Wimbledon Common on 23 June at twelve o'clock. On 25 June, a letter appeared, signed 'A Volunteer', describing how he had crossed Fulham Bridge and arrived at Wimbledon Common to find 5,000 other people there in expectation of a royal ceremony. It seems that the advertisement was a hoax. There is suspicion, but no proof, that the proprietors of Fulham Bridge were responsible for the hoax. Their takings for the day will have been vastly enhanced by the tolls collected from so many people crossing the bridge for an attraction that did not materialise, and having to return disappointed only to find they had to pay the tolls for the second time that day.

At the opposite extreme, when the Thames froze over during the severe winters of 1739–40, 1788–9 and 1813–14, people could cross the river on the ice and avoid the tolls altogether. Frost Fairs were held stretching from Rotherhithe to Putney, with entertainments including bonfires, puppet shows, roundabouts and live animal shows. Even as late as 1870, after the removal of Old London Bridge effectively stopped the Thames freezing in the centre of London, the river was iced over at Putney. Two barges which were stuck in the ice at Hammersmith broke free with a large body of ice, sped through the narrow channel of water in the middle of the frozen river and smashed into one of the wooden piers of Fulham Bridge. The barges both sank and the pier was severely damaged. As pressure was mounting to improve navigation through Fulham Bridge, the proprietors decided to remove the pier and so double the width of the opening at this point.

In 1855, the Chelsea Waterworks Company presented a Bill to Parliament requesting permission to construct an aqueduct over the Thames 100 yards upstream of Fulham Bridge. The purpose was to carry water from its reservoirs in Kingston and on Putney Heath over the river to supply subscribers in west London. The proprietors of Fulham Bridge opposed this Bill on the grounds that it would spoil the beauty of the river by Fulham Bridge and would obstruct navigation. In fact, the bridge by now was looking more and more dilapidated, and the argument about navigation was exactly the same as that used unsuccessfully by the opponents of the original Fulham Bridge Bill. The proprietors went on to suggest that the waterworks company could use Fulham Bridge as a basis for the aqueduct. The objections were not successful, and the aqueduct was completed in 1856. As can be seen from the illustration below, it was a peculiarly ugly structure.

By the middle of the nineteenth century, complaints about the old wooden bridge were growing, both from the shipping companies, which found it difficult to pass through, and the

Fulham Bridge with the aqueduct in 1860

general public, who were met by increasing delays in trying to cross the river as the population of London moved westward. Control of Thames navigation from Staines to the sea had passed from the City of London Corporation to the Thames Conservancy in 1857. After many failed attempts at persuasion, the conservators obtained passage of the Thames Navigation Act in 1870, which gave the Fulham Bridge proprietors powers to borrow money to widen the centre span and also to obtain compensation if the bridge were later subject to compulsory purchase. Eager to save money on their project for widening the centre span, the proprietors were tempted by an offer by John Dixon to use the ironwork from the temporary bridge constructed during the building of the new Blackfriars Bridge. He tendered to widen the arches for £2,800, claiming that the ironwork was still in good condition. However, when the ironwork was examined by a railway engineer, he said it was in fact in very poor condition. The proprietors were forced to turn down the tender and eventually the work was completed in 1872 for £5,492. The new, wider iron central span improved navigation but did nothing to ease congestion and certainly spoiled the quaint appearance of the bridge.

At least two attempts were made to set up companies to build a new bridge to replace the old wooden one. Acts of Parliament were in fact passed, but the companies failed to raise sufficient money. Meanwhile, the Metropolitan Board of Works was set up by the Government, with powers over the improvement of London's transport systems, among other things. The board introduced the Metropolis Toll Bridges Bill in 1877, enabling the purchase of all privately owned bridges in the metropolitan area, with a view to freeing them from tolls. For the first time in their history, all the companies cooperated as they tried to oppose the Bill, but, despite this, it was passed. As described in Chapter 4, the Prince of Wales freed Hammersmith, Fulham and Wandsworth bridges from tolls to wild celebrations on 26 June 1880. This effectively signalled the end of the 150-year-old Fulham Bridge and also the 80-year connection of the Chasmore family with the collection of tolls.

Sir Joseph Bazalgette, as head engineer of the MBW, produced a report in 1880 on the state of all the bridges that had come under his control. Regarding Fulham Bridge, he concluded:

> Scarcely any of the piers are in sound condition; nearly all the piles of which they are formed have been repaired and many of these are now more or less again decayed, so that as a matter of safety it is necessary that early attention should be given to the condition of this bridge.[13]

This damning report signalled the end of Fulham Bridge, as it did for the old Hammersmith Bridge. In the following year, an Act was passed to enable the demolition of Fulham Bridge and the construction of a 44-foot-wide stone bridge in its place. The new bridge, like Hammersmith Bridge, was designed by Sir Joseph Bazalgette, and it was built at a cost of £240,433 by the contractor John Waddell. Its site is where the Chelsea aqueduct crossed the river, and the water mains were incorporated into the bridge, under the pavement. The Prince of Wales laid the memorial stone on 12 July 1884. This can be seen on the west side of the abutment on the Putney side of the bridge. His words expressed a common feeling at the passing of the old in favour of the new industrial age:

> Much as we may regret the disappearance of the bridge with which so many old associates are connected, it is undoubtedly requisite that more extensive facilities should be afforded to render the means of transit across the river easier and safer.[14]

Bazalgette's new Putney Bridge could hardly have looked more different from his ornate Gothic structure at Hammersmith. He chose a classical design, consisting of five segmental arches made of granite from Aberdeen and the Prince of Wales's own quarries in Cornwall. The centre arch is the widest, at 140 feet, and the total width of the river here is 700 feet. John Waddell

Bazalgette's Putney Bridge, opened in 1886

completed the project in four years. To secure the timber piles of the cofferdams, he used steam-powered pile-drivers, which had been invented by James Nasmyth and were first used by him for the construction of the High Level Bridge at Newcastle upon Tyne in 1849.

The bridge was opened as Putney Bridge in May 1886 by the Prince of Wales, who seems to have been extraordinarily active in ceremonies connected with London's Thames bridges. By now, there were few real regrets at the destruction of the old bridge. However, the Bishop of London decided to raise the height of the wall surrounding his garden because the new approach road was much nearer the palace and disturbed his peace. The proprietor of the Eight Bells was displeased for the opposite reason, since the new approach road bypassed his inn. He claimed and received £1,000 compensation from the MBW for the resulting loss of business.

The population of Fulham had increased from about 2,500 in 1729, when the old bridge was built, to well over 40,000 in 1886. Traffic increased considerably over the much more convenient stone bridge, and in 1909 it had to be widened to cater for a tramway. By the time of the 1926 Royal Commission on Cross-river Traffic, Putney Bridge was the busiest bridge to the west of Westminster. The Commission recommended further widening. Unfortunately, All Saints Church stood just to the west of the bridge on the Fulham side, and part of the churchyard was needed for the widening.

All Saints had been the parish church of Fulham since the original church had been built in 1154. In 1880, it had been largely rebuilt by Sir Arthur Blomfield, although the fifteenth-century Kentish ragstone tower was left standing. The churchyard was the burial place of no fewer than eight bishops of London, as well as many local people, before burials ceased there in 1863. The bridge-widening scheme required 15 bodies to be disinterred and moved, much to the displeasure of the vicar. The matter went before the St Paul's Consistory Court, which decided that the Church could not stand in the way of a project that was in the public interest. The bridge was therefore widened to 74 feet at the expense of a small piece of the churchyard and the disturbance of 15 souls, who were reinterred in Fulham Cemetery in Sheen. The tombstones were moved within the churchyard, where they remain today. One of them dates back to 1654.

An equally historic church stands on the Putney side of the bridge. St Mary's has been the parish church of Putney since the thirteenth century. Like All Saints, it was rebuilt in the nineteenth century but retains its fifteenth-century tower. The church is famous for the Putney Debates which took place there in 1647, when Oliver Cromwell and the Parliamentarians set up their headquarters in Putney after Charles I had been confined in Hampton Court. The debates took place around the communion table. The more radical group known as the Levellers proposed such modern concepts as manhood suffrage, freedom of conscience and liberty of the individual. These ideas were opposed by Cromwell and were not implemented when Parliament ran the country after the execution of Charles I in 1649. Unfortunately, St Mary's Church was severely damaged by an arson attack in 1973. An appeal was launched for its restoration. In 1979, on the 250th anniversary of the old wooden bridge, a commemorative medal was struck by the Tower Mint, showing images of the old and new bridges on either side. The bronze medal was sold for £3.50 and the sterling silver medal for £26. The proceeds went towards the successful restoration of the church.

On a quiet Sunday morning, the handsome stone-arched Putney Bridge, with the picturesque parish churches on either bank of the Thames, presents an idyllic picture. It is perhaps surprising that few famous artists or writers have used this scene in their work, whereas the old wooden bridge did inspire a number of writers and artists. Disraeli referred to the 'picturesque bridge' at the end of the King's Road, and the bridge also featured in Charles Dickens' unfinished novel, *The Mystery of Edwin Drood.* The artist Luke Fildes produced an engraving for the front cover of the first edition of *Edwin Drood* showing four of the characters in a boat with Fulham Bridge in the background. James McNeill Whistler also produced an engraving of the bridge just before it was demolished.

The scene, of course, is not always so peaceful. Putney Bridge is still the busiest bridge over the Thames from the south-west, with an average of nearly 60,000 crossings a day during the week. Nor is it peaceful on the day of the Boat Race, when the crowds throng the riverside as they wait for the starting pistol to send the crews on their voyage around the Hammersmith bend and under Hammersmith and Barnes bridges to the finishing line at Mortlake.

Putney Railway Bridge
Until the nineteenth century, Putney had been a relatively small village, consisting mainly of large mansions and estates and market gardens. When, in 1846, the railway came to East Putney Station on the London and South Western line from Nine Elms to Richmond and Windsor, the population changed radically. Almost all the old mansions were gradually demolished and Putney turned into a middle-class suburb. The District Line had

reached Putney Bridge Station on the north bank by 1880, and there was a clear need for the line to be extended across the river to Putney and on to Wimbledon to provide a direct commuting service to the City. Several plans were put forward, but it was the LSWR which finally built the Wimbledon line and the bridge over the Thames. The Act of 1881 stated that the bridge must be of 'ornamental character' and approved by such august bodies as the MBW and the Thames Conservancy. After all the complaints about the ugliness of the recent railway bridge at Charing Cross, Parliament was insistent on having some aesthetic control over future railway bridges.

Mr William Jacomb (1832–87), the LSWR head engineer, designed Putney Railway Bridge and it was built by Head, Wrightson & Co. of Stockton-on-Tees. Work started in 1887 and was completed in March 1889. It consisted of five wrought-iron lattice-girder spans of 153 feet, supported on pairs of cast-iron cylinders which were filled with concrete. An agreement was made to allow the District Line to share the tracks. At first, there were too few trains to provide an adequate commuter service to

Putney Railway Bridge with Putney Bridge in the background

110

Mansion House. However, in 1905 the line was electrified and 108 trains were run each weekday on the District Line. The LSWR service, on the other hand, was never very successful and continued with few trains until it was wound up in 1941. When British Rail took over the railway network after nationalisation in 1948, it also took over ownership of the Wimbledon line and the bridge, even though all the trains were at this stage run by the District Line of London Underground (LU). This strange state of affairs caused controversy when a 600-ton barge rammed into one of the bridge supports in 1991 and caused severe damage. Consequently, a 10-mph speed limit was imposed on all trains running over the bridge. British Rail refused to repair the bridge, as it claimed it could not afford to pay. In 1994, LU took over the ownership of the bridge and railway line. It undertook a 14-million-pound renovation of the bridge and installed a totally new pedestrian walkway. The repairs were completed in January 1998. A plaque on the south end of the bridge commemorates the occasion.

Wandsworth Bridge

The story of Wandsworth Bridge is hardly an illustrious one. Unlike Putney, Wandsworth was never an especially desirable location for aristocrats and the wealthy middle classes. There was no great pressure for a bridge until the late nineteenth century, when the area was dominated by riverside industries around the mouth of the River Wandle, which flows into the Thames just to the north of Young's Ram Brewery. The then fast-flowing Wandle, which meanders with rather less speed today from Croydon to the Thames, was ideal for driving watermills for such industries as flour grinding and calico printing. In 1834, a gasworks was installed in Wandsworth, and by 1864 the riverside area upstream of the Wandle was covered with gasholders. Another large employer was Wandsworth Prison, built in 1851 with no fewer than 708 cells. It had a reputation for a rigorous regime and few managed to escape. It was later famous for incarcerating Oscar Wilde for six months before he was transferred to the less fearsome Reading Gaol, and for the executions of William Joyce (Lord Haw-Haw) and Derek Bentley.

To cater for this increase in the largely working-class population, a company was set up in 1864 by an Act of Parliament to build a bridge there, with the stipulation that it should be 40 feet wide and span the river with no more than three arches. These requirements proved beyond the ability of the company. Consequently, a new Act of 1870 was passed which allowed for a bridge of only 30 feet in width to be carried over the river by five arches. This time, work was started on a design by J.H. Tolmé for a five-span wrought-iron structure. The piers were iron cylinders filled with concrete and sunk 14 feet into the river-bed. The road was carried on iron lattice girders. The bridge should have been completed by early 1873, but the workmen went on strike. Yet another Act was required to allow the company extra time to sort out the strike and finish the bridge. It was finally opened with little ceremony in September 1873. The company had originally hoped that the bridge would allow access to the proposed terminus of the Hammersmith and City Railway on the north bank. Unfortunately, this was never built and profits were meagre.

Following the Metropolis Toll Bridges Act of 1877, Wandsworth Bridge was bought by the MBW for £52,000. This seems generous, since the cost of construction was only £40,000 and the toll income was about half as much as at Putney Bridge, which had been bought for £58,000. J.H. Herring, in his *Thames Bridges from London to Hampton Court* of 1884, wrote a depressing account:

> The new bridge denominated Wandsworth, although the most recent, is perhaps the least frequented by general traffic of any of the metropolitan bridges. Approached by a series of inconvenient roads on the Middlesex side and skirting the few remaining orchards and market gardens of the marshy lands of Fulham, it crosses the Thames to a point on the Surrey side equally inconvenient, midway between Clapham and Wandsworth.

By 1891, the condition of Wandsworth Bridge had deteriorated to such an extent that a 5-ton weight limit was imposed, and in 1927 speed was limited to 10 mph. Many local businesses were thinking of moving because of the poor communications. The 1926 Royal Commission on Cross-river Traffic had recommended that urgent attention be given to Wandsworth Bridge, but the LCC gave priority to work on Chelsea and Putney bridges, much to the annoyance of local people in Wandsworth. Eventually, in 1935, a proposal for a new bridge was put to the Ministry of Transport, which agreed to contribute 60 per cent of the total cost of £503,000.

The LCC engineer, T. Pierson Frank, and architect, E.P. Wheeler, designed a three-span steel cantilever structure with a 60-foot-wide carriageway. The design was presented for approval by the Royal Fine Art Commission with a covering note stating: 'In the design of the bridge a severe simplicity of treatment has been carried out, expressed in a technique essentially related to the material (steel) proposed for its construction.' The eminent and highly cultured members of the Royal Fine Art Commission

Wandsworth Bridge from the newly developed southern river bank

might have been expected to react against any design by engineers and architects from a local authority, especially one constructed in steel. However, they decided to approve the scheme. Doubts were expressed not about the design but about the width of the new bridge. At 60 feet, it was able to cater for only four lanes of traffic, and many thought that six lanes would be required in future. Nevertheless, the work was put out to tender with a specification that all materials should be of British origin or manufacture. The contract was awarded to Messrs Holloway Bros (London) Ltd, which completed the bridge in 1940. Today, it is one of the busiest bridges and carries more than 50,000 vehicles over the Thames per day.

CHAPTER 6

Battersea and Chelsea

Three road bridges and one railway bridge cross the Thames between Chelsea and Battersea. The furthest upstream bridge is the little-known Battersea Railway Bridge of 1863. Battersea Bridge is the first of the road bridges. Opened in 1890, it replaced the old wooden Battersea Bridge of 1771, which was famous for its depiction in James McNeill Whistler's painting *Nocturne: Blue and Gold – Old Battersea Bridge*. Next comes the elegant but now fragile Albert Bridge of 1873, which is the oldest original surviving bridge structure over the Thames downstream of Richmond. Finally, we come to Chelsea Bridge. This rather plain suspension bridge was opened in 1937, replacing a much-admired Victorian suspension bridge of 1858.

Battersea Railway Bridge

This railway bridge crosses the river near the border between Battersea and Wandsworth. It is officially known as Cremorne Bridge after the Cremorne Pleasure Gardens that were located near by on the north bank of the river in the eighteenth century. It may seem a romantic name for a railway bridge, but this is a handsome structure of five curved cast-iron spans resting on four river-piers with stone facings. Unfortunately, not many people have had the opportunity to appreciate its merits, since

Battersea Railway Bridge with the tower of Chelsea Harbour in the background

it is located in what used to be an exclusively industrial area until the recent closure of the Lots Road Power Station and the development of the Chelsea Marina at its northern end. It must be London's least-known river bridge, as it is not easily accessible except by train, and few train services cross it today.

The bridge was built in 1863 by the West London Extension Railway. This company had no rolling stock of its own. All the shares were owned by four of London's railway companies: the London and North Western Railway (LNWR), the Great Western Railway, the London and South Western Railway, and the London, Brighton and South Coast Railway (LBSCR). The bridge was opened on 2 March 1863, on the same day as the new Clapham Junction railway station. It provided a link between Clapham Junction and Kensington that was expected to benefit all four of the shareholding companies. The bridge was designed by William Baker (1817–78), chief engineer of the LNWR. Unfortunately, none of the companies made a success of their services. From 1940 until recently, the bridge was limited to use by freight trains. Today, it is also used by Eurostar trains on their way to the depot in North Kensington and by long-distance services from Brighton to Watford and the North.

Battersea Bridge

Battersea Bridge is the earliest of the bridges between Chelsea and Battersea. Chelsea, which is situated on the north bank of the Thames and was therefore more accessible from the court in Westminster, has a more distinguished history than Battersea. The origin of the name Chelsea is associated with the river. It is thought to come either from the Saxon word 'chesil', meaning 'gravel bank', or the word 'cealchythe', meaning 'landing place for barges to unload cargoes of chalk'. From at least the fourteenth century, there were many wharves for the hiring of wherries and for the use of the wealthy owners of private barges for travelling to and from London. The most famous person to live here was Sir Thomas More, who built himself a mansion near Chelsea Old Church in the sixteenth century. Henry VIII often visited him there for entertainment and conversation, arriving in his royal barge. In 1533, when Henry divorced Catherine of Aragon, More registered his disapproval and the visits ceased. More's last journey from Chelsea was by boat to Lambeth Palace, where he refused to take the Oath of Supremacy which recognised Henry VIII as head of the Church of England and acknowledged its separation from Rome. This resulted in his trial and eventual execution for treason. A modern statue of Sir Thomas More stands today in front of Chelsea Old Church, and the monument he designed for himself and his two wives is located inside the church.

Many distinguished people followed Sir Thomas More in choosing to live in Chelsea, including Sir Hans Sloane, the founder of the British Museum's collections. His descendants developed the area in the eighteenth and nineteenth centuries, when it became associated with famous artists and writers such as Thomas Carlyle, J.M.W. Turner, Dante Gabriel Rosetti, George Eliot and, later, Oscar Wilde. Ranelagh and Cremorne pleasure gardens provided fashionable entertainment for locals and visitors from London. The main local employment was for watermen and boat builders, but there was also work in the Chelsea porcelain factory and in producing the famous Chelsea buns.

On the opposite side of the river, Battersea lay on low, marshy ground often subject to flooding. Hence the soil was fertile and by the eighteenth century became famous for growing asparagus and lavender. The present-day Lavender Hill, which ascends from Clapham Junction Station, reminds us of those days by its name if not by its distinctly suburban atmosphere. A small mixed trading and residential village had grown up around St Mary's Church and the manor house, near which a ferry had crossed the river to Chelsea since the sixteenth century. The ferry is first mentioned in 1550, and it is also listed in a document of 1592 which catalogues all the horse ferries on the Thames in London.

In 1763, Earl Spencer, an ancestor of Princess Diana, bought the manor of Battersea together with ownership of the ferry. He wasted little time in deciding that he could make more money from a toll bridge linking his property to the wealthy inhabitants of Chelsea than from a mere ferry. Earl Spencer formed the Battersea Bridge Company with 16 other investors to raise the finance and manage the maintenance of the bridge with a view to profiting from the tolls. In 1766, an Act was passed authorising the building of a stone bridge along the course of the ferry. However, the estimated cost of £83,000 for its construction was too high, so it was decided to build a wooden crossing instead, for a cost of £15,000. The bridge of 19 narrow arches was designed by Henry Holland (1745–1806) and opened in 1771. It seems that Parliament was not convinced of the reliability of the bridge, as a clause was inserted in the Act to the effect that Earl Spencer must provide a ferry service at the same rate as the bridge tolls in case the bridge had to be closed for repairs.

Unlike Putney Bridge, Battersea was not on a direct route from London, and so returns were low at first. In 1783, the company paid for 50 posters advertising Battersea Bridge as providing the shortest route to Epsom Races. This seems to have had the desired effect, and by 1790 a profit of £1,700 was

1804 view of Old Battersea Bridge with the Horizontal Air Mill
and St Mary's Church spire on the south bank

recorded. Unfortunately, in the severe winter of 1795 the bridge
was damaged by the heavy flow of ice, and no dividends were
paid due to the cost of repairs. In 1799, oil lighting was installed
on Battersea Bridge. This was a first for any bridge on the
Thames. The oil lighting was replaced by gas in 1824. The
Battersea Bridge Company spent much time negotiating with
the Government about payment for the frequent crossings by
troops. In 1818, it was recorded that the Secretary at War, Lord
Palmerston, later Prime Minister, agreed to pay £36 2 s. for that
year's troop crossings. Negotiations were also conducted with
the police, who requested and were granted free passage. This
was clearly a good idea, since Battersea was notorious for
harbouring robbers and vagabonds.

A more serious issue for the company was the proposed new
bridge at Vauxhall, which would have a severe effect on profits.
As always, the company opposed any new bridge. The first
proposal for a bridge at Vauxhall was put forward by none other
than the hapless Ralph Dodd, who, we have seen, was also
involved in the first abortive proposal for Hammersmith Bridge.

The Battersea Bridge proprietors presented a petition to Parliament attacking Dodd's proposal in a very personal manner: 'It is first to be observed that Mr Ralph Dodd, projector of the new bridge, is a well known adventurer and Speculist [*sic*] and the projector of numerous undertakings upon a large scale most if not all of which have failed.' However, when a new Bill was presented and it was clear that Parliament would authorise the bridge at Vauxhall, the company demanded compensation. A clause was inserted in the Act to oblige the Vauxhall Bridge Company to pay each of the Battersea proprietors compensation for loss of revenue following the opening of the new bridge. Vauxhall Bridge was completed in 1816, but no compensation was paid. The matter went to court and in 1821, after a five-year struggle, judgment was finally made in favour of the Battersea proprietors, who received £8,234 compensation.

The Battersea Bridge Company archives throw light on the working conditions of the employees. The following regulations on the terms and conditions under which the tollmen were to work were issued in 1796:

1. That the Salary of each Tollman be twelve shillings a week.

2. That they relieve each other every other day at one quarter before seven o'clock in the morning from the 1st of March to the 30th of September; and from the 1st of October to the 28th of February at a quarter before eight o'clock.

3. That besides the three days and three nights, the Proprietors may require of each of the Tollmen one day's work each week; and upon Sundays both Tollmen to attend at the Hours and Gates as may from time to time be ordered.

4. Not allow any person to drink in the Tollhouse, nor have any liquor there except for their meals.

5. Not to suffer any person to pass without paying Toll, unless the Check Clerk gives permission to do so.

6. If any serious complaint is made against the Tollmen,

the Proprietors for the first offence may fine them five shillings, for the second offence, ten shillings and upon the third offence they must be discharged, as also in case any Peculation or Fraud is proved against them.

7. Each Tollman to be appointed in future must produce a Security for the faithful discharge of his Trust in the sum of Thirty Pounds; and the present Tollmen must take an Oath (as often as the Proprietors may require it) that subsequent to this period they have faithfully accounted for all Moneys they have received, and have never wilfully increased the amount of bad coinage whether halfpence or silver.

8. That provided no complaint is made against them and having complied with the above Laws and Regulations, they shall be paid a Gratuity every month of five shillings each.[15]

In 1810 the company raised the tollmen's salaries to 16 shillings a week and stipulated 'that three pints of Porter be allowed them every day they are on duty except that the allowance be stopped on the days that any Person is found tippling in the Tollhouse'. Shortly after this, the salary was increased to 18 shillings and the beer allowance stopped.

By the second half of the nineteenth century, the bridge had undergone so many repairs that it looked like a dilapidated ancient monument. Its safety record was poor, it was clearly inadequate for the increasing traffic that resulted from the growing population of both Chelsea and Battersea, and it was an obstruction to shipping. Opinion was divided on its aesthetic merits. George Bryan wrote in 1869:

The number of lives that have been sacrificed at this bridge, together with the barges sunken at it even within the last few years is really too painful to contemplate. It is a sad contrast in every respect to the elegant structures that now span the river and it is to be hoped there will soon be erected another one in its place.[16]

121

On the other hand, many artists produced paintings and engravings which tended to show the bridge as a picturesque and even romantic structure. Walter Greaves, whose most famous painting was of Hammersmith Bridge on Boat Race day, worked in the family firm of boat builders close to Battersea Bridge. He was evidently fascinated by the swift flow of the water through the narrow arches, where young watermen took up the challenge to shoot the bridge at considerable risk to themselves. He recalled how as a boy he once watched a barge sink when it collided with one of the piers, and local watermen dragged the corpses of an entire family from its cabin. Greaves frequently depicted the bridge and the adjacent riverside in a rough but naturalistic style.

Whistler, who had been Greaves's artistic mentor but who later fell out with him, used a very different approach in his series of nocturnes. He depicted a variety of riverside scenes including some at Chelsea. Like Greaves, Whistler had a house in Chelsea, and the area by Battersea Bridge was familiar ground to him. His *Nocturne: Blue and Gold*, painted between 1872 and 1875, shows Battersea Bridge looming out of the water in a gloomy haze and lacking any clear details. It reminds us of a late Turner or even a Monet. Whistler's nocturnes achieved notoriety when John Ruskin wrote a scathing attack on them, accusing the artist of 'flinging a pot of paint in the public's face' with his *Nocturne in Black and Gold: The Falling Rocket*. Whistler sued for libel and won the case. The judge, however, seems not to have taken the matter too seriously. Referring to *Nocturne: Blue and Gold*, he caused much laughter in court when he asked Whistler, 'Which part of the picture is the bridge?', and he awarded damages of just one farthing. The painting was startling, even revolutionary, at the time and now hangs in Tate Britain. The bridge itself did not survive so long. However, its artistic memory is preserved by the statue of Whistler, sculpted by Nicholas Dimbleby in 2005, which stands at the north end of today's Battersea Bridge.

In 1864, the Albert Bridge Company obtained an Act of Parliament enabling them to build a bridge a few hundred yards downstream of Battersea Bridge. As this would seriously damage the finances of the Battersea proprietors, a clause was inserted to give powers of compulsory purchase, so that the new company took over ownership of Battersea Bridge before starting work on the Albert Bridge. In 1879, the Metropolitan Board of Works bought both bridges for £170,000 and freed them from tolls. Soon after this, Battersea Bridge was declared unsafe. In 1883, after 111 years of constant repairs, it was closed and replaced by a temporary footbridge while a new Battersea Bridge was constructed to the design of Sir Joseph Bazalgette.

Bazalgette's bridge consists of five segmental arched spans constructed of cast iron. It is one of the narrowest of London's Thames bridges, with a roadway width of only 40 feet. The two 8-foot-wide footpaths are supported on arms cantilevered out from the main structure of the bridge. Although not as ornamental as Bazalgette's Hammersmith Bridge, his Battersea Bridge has some fine cast-iron decoration. The parapet consists of a line of delicate arabesque arches. The main spans are

Bazalgette's Battersea Bridge of 1890, viewed by Derwent Wood's statue of Atalanta on the Chelsea Embankment

topped by a strong cornice and have ornamental shields and foliate decoration in the spandrels. The bridge was opened on 31 July 1890 by Lord Rosebery, the first chairman of the London County Council. Since almost all previous metropolitan bridge-opening ceremonies had been conducted by a member of the Royal Family, he decided to meet any objections head-on in his opening address, as reported in *The Times* of 22 July 1890: 'I understand that there are some in this neighbourhood who make it a matter of complaint that I am not of Royal extraction. (Laughter.) That is a feeling which I entirely share. (Renewed laughter.)' This experience must have given him great confidence for his next opening ceremony, when on the following day he opened the major new road between Clerkenwell and Holborn, and named it Rosebery Avenue after himself.

Although the new Battersea Bridge was praised by Lord Rosebery at its opening, like many of its Victorian counterparts it was designed before the invention of the motor car and has proved less than satisfactory for modern traffic. It has also caused problems for navigation, as it is low and the swirling current on the bend in the river makes steering difficult. Passing boats have crashed into the bridge on a number of occasions, often resulting in its closure for repairs. In March 1950, the collier *John Hopkinson* caused severe damage to the centre pier when it was blown into it during a gale. The LCC was concerned that the whole structure would collapse and closed the bridge for repairs until January 1951. Most of the cost of the repairs was borne by the local ratepayers because the Merchant Shipping Act limits the liability of vessel owners to £8 per ton. Since the *John Hopkinson* weighed 1,300 tons, the owners had to pay only £10,400 out of the total of £35,000. The reaction of local residents when the bridge was eventually reopened was remarkable. The *Star* of 17 January 1951 quotes one resident as saying, 'The return of the traffic noise makes us feel we are a community again. It was really ghostly, especially at weekends – almost like living in the country.' The point was also made that local traders estimated they were losing £15 a day on average during the closure.

The most serious incident from a human point of view involved the motor vessel *Delta*, which became jammed under the bridge in 1948. Its master, Hendrikus Oostring, was trapped in the wreckage of the wheelhouse for some time before he could be rescued. He had to be operated on for broken arms and the *Delta* itself was towed away to dry dock for repairs. On this occasion, Battersea Bridge suffered only minor damage. More recently, on 21 September 2005, a barge crashed into the bridge causing its immediate closure. The following day, pedestrians and cyclists were allowed to cross, but the bridge remained closed to vehicle traffic for several months while repairs were undertaken.

Battersea Bridge is now Grade II listed and currently there are no plans to replace the Victorian structure. Its history is relatively uneventful apart from its frequent closures. *The Times* of 26 May 1958 does record an amusing incident when five youths jumped naked off the bridge at midnight after a heavy bout of drinking. They had to cling to a boat for several minutes before the police came to rescue them. They were immediately taken to Battersea Police Station wearing just their underpants and charged with being drunk and disorderly.

On a sadder note, Battersea Bridge witnessed the furthest limit of the fatal journey of a 19-foot-long northern bottlenose whale which swam 40 miles up the Thames on 20 and 21 January 2006. Normally, these rare mammals inhabit the Atlantic Ocean, and it is thought that the whale must have taken a wrong turn off the north of Scotland and so ended up at the Thames estuary. There, its natural instinct to travel west so as to return to its usual habitat caused it to swim up the Thames. Attempts were made to persuade the whale to swim back downstream to the North Sea, but to no avail. Eventually, rescuers managed to lift the massive bulk onto a barge, which then proceeded on its way towards the sea. Unfortunately, the whale was by then exhausted from lack of food. It died before it could be returned to the open sea, but not before it had created headline news throughout the world for its 48-hour battle for survival in the River Thames.

Albert Bridge

The Albert Bridge crosses the Thames from a little to the east of Chelsea Old Church to the west side of Battersea Park. It is named after Prince Albert, who had the original idea for a bridge here. Although Albert died before the bridge was completed, it still bears his name today, unlike the next bridge downstream, which was opened as Victoria Bridge but has since changed its name to Chelsea Bridge. The Albert Bridge proposal was fiercely contested by the Battersea Bridge proprietors. It was finally allowed to go ahead when the 1864 Act of Parliament forced the Albert Bridge Company to take over ownership of the by-now-decrepit Battersea Bridge under a compulsory purchase order. The Act specified that the bridge should be completed within five years and also laid down the rates of the tolls.

Construction was delayed because of uncertainty about the plans to build an embankment at Chelsea. Consequently, a new Act was required to allow an extension of the time limit. Work started in 1870 and the bridge was opened without ceremony in 1873 for a total cost of £200,000. The design of the bridge, by Rowland Mason Ordish (1824–86), was innovative and almost unique. The two river-piers rest on cast-iron cylinders filled with concrete. These were the largest castings ever made at the time. The roadway is supported both by cable-stayed rods, which radiate out from the top of the twin ornamental cast-iron towers resting on each of the piers, and also by suspension chains. Because the towers were constructed outside the parapets, they do not obstruct the pathways.

The only other bridge to use Ordish's system was the old Franz Joseph Bridge in Prague, which was over 800 feet long, compared with the 710 feet of the Albert Bridge. The Franz Joseph Bridge was demolished in 1949, while the Albert Bridge still stands today as the only original vehicular bridge structure remaining on the Thames in central London, apart from Tower Bridge. It is perhaps surprising that cable-stayed designs were hardly used again until the latter half of the twentieth century. Today, some of the most imposing bridges in the world, including the Queen Elizabeth II Bridge at Dartford, with its 450-metre main span, are cable-stayed.

Like Bazalgette's Battersea Bridge, Albert Bridge has not worn well and, with its narrow carriageway of 27 feet, has proved unsatisfactory for modern traffic. Problems were first noticed when the bridge was taken over by the MBW and Sir Joseph Bazalgette diagnosed rust in the supporting chains. New chains were installed in 1887. In the 1950s, the LCC planned to demolish the old bridge and build a new, wider one. John Betjeman led the campaign to save the old Albert Bridge, which became known as 'the Trembling Lady' because of its tendency to wobble, especially when troops from the nearby Chelsea Barracks marched across. *The Times* published several letters both supporting and opposing the LCC plans. Betjeman used the full power of his invective against a scientist, Mr Hill, who had ventured to suggest that we should move with the times and build for the twentieth century. In his letter of 9 June 1957, Betjeman wrote:

> I quote the statement of your correspondent of June 7 that it is a sign of progress when a nation destroys the buildings of the days of its greatness. I cannot believe he wants the country to be all airports, glass towers and sodium lights, with no old bridges at all.

Unlike the abortive campaign to save Waterloo Bridge in the 1930s, this protest was successful, and the LCC abandoned its plans at least for a while.

By the 1970s, the bridge had become so fragile that a weight limit of 2 tons was imposed. The LCC produced another plan, this time to close the bridge to traffic and limit it to pedestrians and cyclists. This plan again aroused fierce controversy, and supporters and opponents enlisted the endorsement of celebrities. The Royal Automobile Club, which opposed the idea, brought in the film star Diana Dors to front its campaign. Local groups had the support of the poet Robert Graves in their campaign to close the bridge. The latter group managed to

127

obtain 1,000 signatures for a petition, but this was derided by the British Road Federation, which wrote: 'If you send a lot of students around to council flats, most people will sign anything without knowing what it is all about.' The 1974 public inquiry decided in the end not to close the bridge to traffic, although the 2-ton weight limit was maintained and work was undertaken to strengthen the deck. More controversially, the LCC was allowed to install two new cylindrical concrete piers to support the middle of the deck despite protests from the Royal Fine Art Commission, which argued that this would entirely spoil the elegance of Ordish's design. These piers were supposed to be a temporary measure to give the LCC five years to decide what to do with the bridge. The chief engineer of the LCC had told the public inquiry that the bridge could last a maximum of 30 years even with the central supports. In the event, the supports were never removed and the bridge still stands.

Albert Bridge is now Grade II listed. Its survival would have pleased William Blake, who lived nearby and admired it greatly.

Albert Bridge with night-time reflections in the river

During the 1990s, it underwent considerable redecoration and rewiring, which has resulted in making the bridge one of London's most attractive sights both in the daytime and at night. The delicate Wedgwood colour scheme of salmon pink, light and dark blues and green complement the ornamental intricacy of its cast-iron towers and the slim wrought-iron bars splayed out from their tops. The engineers claim that the colour scheme is also designed to stand out in murky light, so that shipping will not collide with the piers. The bridge really comes to life at night, with its award-winning lighting system. Four thousand low-voltage tungsten halogen bulbs hang from its cables and towers, lighting up the river in a glitter of fairyland. As well as adorning the river at night, the bulbs have a lifespan of 18 months, much longer than traditional lighting, and so the bridge does not need to be closed too frequently for changing the bulbs. A Certificate of Commendation was awarded in 1993

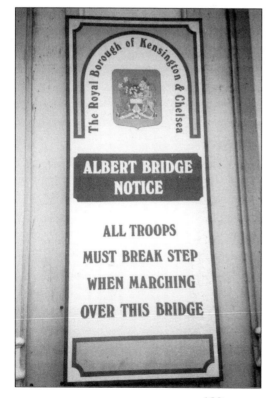

The notice to troops to break step at the approach to Albert Bridge

by the Lighting Industry Federation for this innovative lighting system and was presented by Mary Archer, chair of the National Energy Council. Albert Bridge has come of age in style, hopefully to last many more years, provided that traffic obeys the 2-ton weight limit, that troops obey the notice at its approach requesting them 'to break step when crossing the bridge' and that the central props do not have to be removed, despite the fact that they were supposed to be only a temporary measure 30 years ago.

Chelsea Bridge

The first Chelsea Bridge was closely connected with government plans to develop the unruly Battersea Fields area into a pleasure park for public enjoyment. In 1842, the Commission for Improving the Metropolis recommended that a suspension bridge be constructed between a point downstream of the Royal Hospital Chelsea and the notorious Red House Inn on the south bank so as to allow access to the proposed Battersea Park.

The Red House had been a popular destination for revellers since the sixteenth century, when people travelled there by wherry from London. One moralist described the area around the inn as a place where 'lawlessness, Sabbath desecration, immorality and vice are rampant'.[17] Sundays, when fairs were held with pugilistic fights, dog fights and general rowdiness, were especially riotous. The inn was also notorious for duels, which were officially illegal but often took place there because of the lack of police presence. One famous duel occurred when the Duke of Wellington, hero of Waterloo and Prime Minister, challenged the young Earl of Winchilsea, who had insulted him because of his support for the Catholic Emancipation Bill in 1829. With their seconds present, the two men walked to their places. The Duke fired first and missed. It is not known if this was deliberate, but he insisted that Winchilsea fire in his turn. Winchilsea fired into the air and rendered his apology, thus ending the matter in a gentlemanly manner.

In 1846, an Act of Parliament was passed authorising Her Majesty's Commissioners of Woods and Forests to purchase the

130

Red House Inn and some 200 acres of the surrounding land with a view to developing it into a park and to building an approach road from Sloane Square and a bridge over the river. Work could not start on the new Battersea Park until the Chelsea Waterworks Company, which occupied the site of today's Grosvenor Canal, was obliged to move up river because of the 1852 law forbidding water to be taken from the Thames below Teddington. Until then, the company had filled its reservoirs in Hyde Park using elm trunks as pipes to take water from near where the River Westbourne flowed into the Thames. The Westbourne had become heavily polluted, and Tobias Smollett described the water taken from the Thames here as

> impregnated with all the filth of London and Westminster
> – human excrement is the least offensive part of the
> concrete, which is composed of all the drugs, minerals
> and poisons used in mechanics and manufacture,
> enriched with the putrefying carcasses of beasts and men,
> and mixed with the scourings of all the washtubs, kennels
> and common sewers within the bills of mortality.[18]

Shortly afterwards, the Westbourne was culverted and made into a sewer. The new bridge and park were about to provide healthy exercise for the public and at the same time remove a cholera-bearing water supply and a place of immorality. The Commission for Improving the Metropolis had done its work well. Unfortunately, the Act had allowed for tolls to be charged even though the bridge was in public ownership.

In 1857, while excavating the river-bed for the foundations of the river-piers, workmen came across one of the most exciting Iron Age finds in Britain. The Battersea Shield was made of bronze, with highly embossed roundels containing 27 embedded circles of red enamel. It is of exquisite workmanship and belies the idea that our prehistoric ancestors were primitive philistines as regards art and craftsmanship. Today, it can be seen in the British Museum. Other objects of Ancient British and Roman origin were found in the same area and it was originally thought that there must have been a battle here between the Roman invaders and the British tribes. This seemed even more likely as this is one of the places where Julius Caesar may have crossed the Thames in 55 BC. The river must have been fordable at that time, as even in 1948 it was recorded in *The Times* that a Mr Joe Simmons walked across at low tide, and the level of the river would have been lower in Roman times. However, the Battersea Shield was never used in battle, nor was it made for real fighting. It is now thought that it was a ceremonial offering, as it was common to throw valuable items into the river to honour the gods.

Although the Act authorising Chelsea Bridge was passed in 1846, progress was slow and this caused much anger and derision. This is typified by a letter in *The Times* of 25 December 1856 signed by Pons Asinorum (Bridge of Asses). The writer contrasted the excellent efforts made by the Commissioner of Works, Sir Benjamin Hall, in the huge project of laying out the new Battersea Park with his dilatory approach to constructing the bridge. Sir Benjamin may have been distracted by his responsibilities for Big Ben, which was being built at the same time and is named after him.

Chelsea Bridge was of a conventional suspension type. It was designed by Thomas Page, who later designed the replacement Westminster Bridge. The cast-iron towers rose 97 feet above Thames High Water to support the wrought-iron suspension chains which carried the wrought-iron girders of the roadway. The ironwork was constructed in Edinburgh and erected in 1857. It was inscribed with the date of construction and the words 'Gloria Deo in Excelsis' on its 347-foot central span. Queen Victoria opened the bridge on 28 March 1858 and named it the Victoria Bridge. It was described by the *Illustrated London News* of 25 September 1858 as 'a fairy structure, with its beautiful towers, gilded and painted to resemble light coloured bronze, and crowned with globular lamps, diffusing light all around'. The unpopular tolls were collected from picturesque octagonal lodges at either end of the bridge.

The opening of Old Chelsea Bridge in March 1858

❖ ❖ ❖ ❖ ❖ ❖ ❖

Thomas Page (1803–77)

Thomas Page was born in London but educated on Teesside, where he trained for a career at sea. However, he decided to join an engineering works in Leeds, before entering the employment of the architect Edward Blore, for whom he worked as a draughtsman. The combination of engineering and architectural experience provided a firm basis for his future career as a bridge designer. His first major project was the construction of the Thames Tunnel, for which he was the engineering assistant to Isambard Kingdom Brunel. The project took twenty years to complete, and during this time there were five inundations, resulting in considerable loss of life.

Because of his practical knowledge of the Thames river-bed, Page was asked to give evidence to the many select committees on Westminster Bridge, reporting on its state of repair. Partly as a result of his success at Chelsea Bridge, it was his design that was finally accepted in 1854 when the decision had been taken to rebuild Westminster Bridge. He later designed several bridges in England, including a bridge at Datchet over the upper Thames. In 1870, towards the end of his career, he proposed a submerged tube tunnel under the English Channel, predating the Channel Tunnel by over a century. This proposal was certainly far in advance of the currently available technology and fortunately was not pursued further.

On 4 July 1857, a massive demonstration was held by 6,000 people to protest against the tolls, which would stop the poorer classes enjoying the park. Shortly after the bridge was opened in the following year, the tolls were abolished on Sundays and statutory holidays, which were about the only times most people could visit the park. The park itself was designed by the architect James Pennethorne, based on the original idea of Thomas Cubitt, the developer of Belgravia. In order to build his elegant stuccoed housing in Belgravia, Cubitt had used the excavated spoil from St Katherine's Docks to fill in the marshy land; and for the equally marshy land of Battersea Park, spoil was brought from the concurrent excavation of the massive Royal Victoria Dock. By 1865, 50,000 visitors a year came to enjoy the amenities of Battersea Park, where exotic plants such as palm trees grew in the open, and wholesome eateries and tennis courts had replaced the rowdy fairs of the Red House days.

Unlike the neighbouring Albert Bridge, the Victoria Bridge soon changed its name, and is now known as Chelsea Bridge. It seems that the name was changed because the authorities were concerned about its long-term safety and did not want its possible collapse to be associated with Queen Victoria. Like the Albert and Hammersmith suspension-type bridges, Chelsea Bridge suffered from continual problems of deterioration as well as being too narrow for motor traffic, which doubled from 6,000 per day in 1914 to 12,000 per day in 1929. Even its once-admired architectural style was challenged by Sir Reginald Blomfield, one-time president of the Royal Institute of British Architects (RIBA) and designer of Lambeth Bridge. In 1921, he wrote of Chelsea Bridge:

Its kiosques and gilt finials, its travesty of Gothic
architecture in cast iron, its bad construction and its text
of 'Gloria Deo in Excelsis' above the arch between the
piers, are redolent of 1851, the year of the Great
Exhibition, the *locus classicus* of bad art, false enthusiasms
and shams.[19]

Victoriana had by then gone out of fashion and was not to return
to favour until well into the latter half of the century. The gilt
finials on the tops of the towers had to be removed in 1922 when
they were found to be unsafe. In 1926, the Royal Commission on
Cross-river Traffic included Chelsea among the bridges that
should be rebuilt.

In 1931, the LCC decided that the old bridge should be
replaced by a new six-lane suspension bridge, costing £695,000.
Unfortunately, this was a time of austerity, and the MOT
overruled the LCC, specifying that only four lanes should be
provided. In 1933, the LCC agreed to go ahead with a

Chelsea Bridge of 1937, with its galleon lamp-posts

suspension bridge carrying four lanes at a cost of £365,000 provided that the MOT contributed 60 per cent of the cost. At the time, this must have seemed a bargain. Today, we may regret that the bridge was not built wider, since the cost of adding two extra lanes now would be considerably more than £330,000, even allowing for inflation.

The overall design of the bridge was by the LCC architects G. Topham Forrest and E.P. Wheeler and was approved by the Royal Fine Art Commission. In contrast to the highly ornamental Victorian structure, the new Chelsea Bridge was distinguished by its restrained simplicity. The reason for this was said to be that the view was blocked by the nearby Grosvenor Railway Bridge and by the massive Battersea Power Station, designed by Giles Gilbert Scott, who was also to design the controversial replacement Waterloo Bridge two years later. The only embellishments are the lamp-posts, decorated with golden galleons, and the coats of arms of the LCC, Battersea, and Kensington and Chelsea boroughs, which are at each end of the bridge. The light bulbs which hang from the suspension chains and the carriageway make it a worthy companion to the neighbouring Albert Bridge at night-time.

Rendel, Palmer and Triton supervised the engineering of the bridge and the construction contract was performed by Messrs Holloway Bros (London) Ltd. Chelsea Bridge is built of steel with granite-faced river-piers. Its overall suspended length is 698 feet with a 322-foot central span and with 24 feet of headroom above Trinity High Water. It was specified that all the materials used for the construction of the bridge should be sourced from the British Empire. The wood paving which supported the carriageway was laid with Douglas fir from British Columbia in Canada, and so it was appropriate that it was opened by W.L. Mackenzie King, the Prime Minister of Canada. The ceremony took place on 7 May 1937, five days before the coronation of George VI, which the premier was due to attend.

The Second World War started two years after the bridge was opened. Since the Chelsea Barracks were situated on the north of the river near the Royal Hospital Chelsea, the Royal

Engineers constructed a temporary bridge in the vicinity in case any of the local bridges were destroyed. No enemy action took place in this area during the war, and the temporary bridge was removed in 1945. One battle, however, did take place on the bridge in 1971. This involved a clash of two rival motorcycle gangs, the Essex Angels and the Road Rats. About fifty men were present and of them twenty were arrested and sentenced to between one and twelve years in prison. As recorded in the *Daily Telegraph* of 3 March 1971, weapons used included iron bars, knives, motorcycle chains, shotguns and even one medieval ball and mace with spikes. The Battersea Shield would have offered no protection against this armoury.

CHAPTER 7

Vauxhall and Lambeth

Three bridges cross the Thames between Chelsea and Westminster. Grosvenor Railway Bridge is located only 200 yards to the east of Chelsea Bridge. It is a railway bridge, constructed originally in 1860 but considerably extended since then. Vauxhall Bridge is located about 1,000 yards downstream and crosses the river between the MI6 building on the south bank and Tate Britain on the north bank. It was opened in 1906, when it replaced the first iron bridge to be built over London's river. Lambeth Bridge, which was built in the 1930s to replace a suspension bridge of 1862, is a further 500 yards downstream.

Grosvenor Railway Bridge

This bridge carries the services of today's Southern, South West Trains and South Eastern Trains railway companies across the river to Victoria Station. The history of Victoria Station and the railway companies, both private and nationalised, that have used it is highly complex and cannot be covered here in detail. The initiative which resulted in bringing rail services to Victoria came from the West End and Crystal Palace Railway Company, which had been formed to take advantage of the expected increase in the number of people wanting to travel to Sydenham to visit the Crystal Palace. This wondrous glass structure had been moved there from Hyde Park after the Great Exhibition of 1851. Initially, the railway was only extended as far as a station called Pimlico, which was built on the south bank of the river and opened in 1858 on the same day as the first Chelsea Bridge.

The London, Brighton and South Coast Railway (LBSCR) soon took over the service. It was clear that a far better service could be provided by taking trains across the river to nearer the West End rather than leaving them at Pimlico Station in a desolate and inaccessible area on the south bank. On 23 July 1858, the company obtained authorisation to build a railway bridge and to extend the line across the river to a site near the Grosvenor Canal, which had originally been constructed by the now-defunct Chelsea Waterworks Company. Since the LBSCR could not justify the investment on its own, it brought in two other companies, the London, Chatham and Dover Railway (LCDR) and the Great Western Railway, which subscribed 50 per cent of the cost between them. The GWR used the broad-gauge lines measuring 7 feet ¼ inch, whereas all other companies had standardised on lines of 4 feet 8 ½ inches, so a mixed gauge had to be installed for the two railway tracks carried by the original bridge.

Grosvenor Railway Bridge, which was the first railway bridge to cross the Thames in the central London area, was designed by Sir John Fowler and consisted of four 175-foot wrought-iron spans. It was built remarkably quickly, as the first stone was laid on 9 June 1859 and the first train crossed to the new Victoria Station exactly one year later on 9 June 1860. Although sometimes called Victoria Bridge because it leads to Victoria Station, the bridge's real name is Grosvenor, the family name of the Duke of Westminster, who owned the land on which the railway line and station were built and who had recently developed the Belgravia and Pimlico estates nearby.

Sir John Fowler (1817–98)

John Fowler was born in Sheffield and worked on several railway projects in the north of England before setting up his own practice in Westminster in 1844. He won the contract for building the Severn Valley Railway when he designed two 200-foot cast-iron bridges to span the River Severn. His first London project was the design of Victoria Station and the Grosvenor Railway Bridge for the LBSCR. A number of other London railway projects followed, during which he worked with Benjamin Baker, who became his partner in 1875. This partnership resulted in one of the greatest achievements of the Victorian Age – the construction of the Forth Railway Bridge. Fowler had already been knighted for services in Egypt and the Sudan, where his map surveys had been used by Kitchener during the relief of Khartoum. Following the success of the Forth Bridge project, he became a baronet, while Baker received a knighthood. They continued to work together and in 1890 they constructed the world's first deep-level Underground line using a tunnelling machine invented by J.H. Greathead.

Many changes have occurred on the railway lines to Victoria since the Grosvenor Railway Bridge was first opened, resulting in much widening and reconstruction. The first major works on the bridge were undertaken in 1866. By this time the GWR no longer used the tracks and so the five new lines constructed at that time were single gauge. The necessary widening made the Grosvenor the widest railway bridge in the world at the time. In 1901, the bridge was again widened to take two extra tracks, making nine in all. In 1909, London's first electric train crossed the Grosvenor Railway Bridge on the service from Victoria to London Bridge. The trains used a system of overhead current collection which was not replaced by the normal third-rail system until 1928.

By 1963, the bridge had deteriorated to such an extent that it had to be completely rebuilt. Fortunately, the old bridge had been constructed at several different times and was in effect three separate bridges joined together. This allowed the tracks of two of the bridges to remain open throughout the project.

Grosvenor Railway Bridge with Battersea Power Station in the background

The bridge has been reconstructed as ten separate bridges joined together to carry ten railway tracks over four steel spans. The new bridge was completed in 1965. Today, it carries by far the largest number of trains across the river of any of London's railway bridges for the wide variety of services to Victoria Station.

In the past, Victoria Station was the starting point for thousands of passengers who travelled to France and beyond on the boat trains. This was reflected by the inscription on the front of Victoria Station's entrance arch, 'The Gateway to the Continent', which has now been removed. The war memorial there also reminds us of the many trains that transported troops to France during the First World War. On a happier note, Victoria Station has seen the arrival of more royalty and visiting heads of state than any other of London's termini. Today, the station is the second-busiest London terminus after Waterloo, and commuters, jammed together in crowded compartments, may not be aware of its glamorous past as the trains slow down at the approach via Grosvenor Railway Bridge.

Instead, there is usually time to admire the looming presence of the now-derelict Battersea Power Station at the southern approach to the bridge. The power station was designed by Sir Giles Gilbert Scott, who also designed Waterloo Bridge. It was opened in 1933, extended to its present size by the addition of two extra chimneys in 1953, and finally closed in 1983. It has been derelict since then. The site has now been purchased by Parkview International (London) PLC, which has obtained planning permission for an ambitious new mixed-use development, including a single-table restaurant on top of one of the chimneys. Passengers crossing Grosvenor Railway Bridge also have a fine view to the west of the adjacent Chelsea Bridge and, beyond that, Albert Bridge. Together, they present a magical picture, especially when lit up at night. To the east, Vauxhall Bridge is just out of sight, as it is located round the steep bend in the river at Vauxhall. On the north side of the bridge is the now mainly filled-in Grosvenor Canal, with the tall chimney of the Western Pumping Station.

The pumping station was built in 1875 as part of Sir Joseph Bazalgette's London sewage and drainage system, and is still in use today. During one especially heavy storm in the autumn of 2004, which overloaded the Victorian sewage pipes, it pumped tons of raw sewage into the Thames. According to a BBC news report, this resulted in a severe reduction of the level of oxygen in the water and caused the death of an estimated two million fish over a period of two days. A permanent solution, involving the construction of new intercepting sewage tunnels, would cost billions. Therefore Thames Water will continue the stopgap solution of using their 'bubble boats' to pump oxygen back into the river whenever storm pollution occurs. Despite its earthy origins, the pumping station is now a listed building and a fitting companion to the bridge, which, when originally built, predated it by 15 years.

Vauxhall Bridge

Vauxhall and Lambeth have been closely linked for centuries and today are both controlled by the London Borough of Lambeth. The first mention of Lambeth appears in a charter granting the manor to the monks of Waltham Abbey in 1062. In 1197, the manor came into the hands of the Archbishop of Canterbury, who, in exchange, gave the current owner, the Bishop of Rochester, some lands in Kent. Since then, Lambeth has been the London residence of the archbishops of Canterbury, who built and extended Lambeth Palace in order to keep close to the monarch and the seat of government across the river in Westminster. Until the seventeenth century, Lambeth Palace stood alone on the marshy south bank of the Thames and few buildings were to be found between it and Southwark to the east or Battersea to the west.

One man who did have a house in Lambeth was Robert Catesby. In 1605, he used his house to store the gunpowder which, with Guy Fawkes and his other co-conspirators, he took across the river to deposit in the cellars under the House of Lords with the aim of blowing up Parliament and installing a Roman Catholic government. One of the conspirators warned his brother-in-law, Lord Monteagle, not to attend the House of Lords on 5 November. Monteagle informed the Government, and the gunpowder plot collapsed when Guy Fawkes, together with several barrels of gunpowder, was discovered that night in the House of Lords cellars.

As London expanded westwards towards the end of the seventeenth century, various industries were set up on the south bank at Lambeth to service the increasingly affluent inhabitants of Westminster. Venetian glass artists and Dutch potters were among the first to arrive. In 1815, John Doulton, who had been apprenticed to John Dwight's Fulham Pottery, set up his factory on the riverside. Soon the firm of Doulton and Watts was producing sanitary ware including stoneware pipes which disgorged vast quantities of sewage into the Thames and thus contributed to its pollution. By this time, the green fields around Lambeth Palace were being replaced by a whole range of industrial enterprises belching forth fumes and shattering nature's calm with the cacophony of the Industrial Revolution. This resulted in a growing, mainly working-class, population, housed in tenements in narrow streets.

Amazingly, next to all this industry a number of pleasure gardens were developed. The most famous was the Vauxhall Gardens, located in Nine Elms on part of the site of today's New Covent Garden Market. The site was to the west of the developed area, and these gardens continued to function until 1859. However, the smaller Cumberland Gardens did not survive so long. They were near the end of the Vauxhall Turnpike, which ran to the south of the river, and this was the most convenient place for building a bridge to provide a crossing to the growing neighbourhoods of Pimlico, Belgravia and Knightsbridge on the north bank.

Recent excavations of the river-bed at Vauxhall by the Museum of London Archaeology Society (MoLAS) have established that a bridge or at least a long jetty existed here in prehistoric times. In 1993, remains of substantial wooden piled foundations were found to the west of today's Vauxhall Bridge. The sand and silt had been washed away to reveal their blackened tops, and today they can be seen emerging from the south side of the river-bed at low tide. Radiocarbon measurements were taken and these dated the structure to no later than 300 BC, making this much the earliest bridge over the Thames. Since no mention was made by Julius Caesar of such a bridge when he crossed the river during his conquest of Britain, we must assume that it had not survived or had been destroyed by that time.

The first proposal for a bridge at Vauxhall in historic times was put forward in 1806. At that time, the nearest bridges were a mile downstream at Westminster and nearly two miles upstream at Battersea. Apart from these bridges, the horse ferry at Lambeth provided a permanent service from near Lambeth Palace to the site of today's Horseferry Road, and there was a Sunday ferry for visitors to Vauxhall Gardens.

The 1806 proposal for a bridge at Vauxhall was sponsored by Ralph Dodd. This was successfully opposed by the proprietors of Battersea Bridge, as described in Chapter 6. Another proposal in 1807 also failed, but in 1809 an Act was finally passed incorporating the Vauxhall Bridge Company with the authority to 'build a bridge across the Thames from the bank or shore thereof at or near a certain place on the south side of the said river called Cumberland Gardens, near Vauxhall Turnpike, to the opposite shore called Millbank'. Millbank was hardly more salubrious than Lambeth at the time. Like Lambeth, the area was marshy and although close to Westminster, it was still rural to the south of Horseferry Road. Part of the land was being bought up in order to construct the massive Millbank Penitentiary, which was to become one of the most unhealthy of London's prisons. However, further to the north lay the prosperous and fashionable developments at Knightsbridge and Belgravia; a bridge here would allow more direct access to the West End from the south than that provided by Battersea Bridge. The Act stipulated that the Battersea proprietors should continue to receive tolls at a level as if the new bridge did not exist. As described in Chapter 5, this led to a lengthy court case when the Vauxhall Bridge Company failed to pay adequate compensation.

The Act of 1809 enabled the company to raise up to £300,000 by the sale of shares or by a mortgage and to take all the profits from the tolls. The tolls were set at between two shillings and sixpence for vehicles drawn by six horses and one penny for pedestrians. There were exemptions for mail coaches, soldiers on duty and parliamentary candidates during election campaigns. The company was obliged to pay compensation not only to the Battersea proprietors but also to the watermen who ran the Sunday Vauxhall Gardens ferry. If agreement could not be reached, the amount was to be decided by a jury of '24 honest, sufficient and indifferent men'. Such men may have been hard to find given the level of corruption prevalent in public life at the time. Anyone convicted of damaging the bridge had to pay a fine of 40 shillings, and if they failed to do so, would be committed 'to the nearest Bridewell or House of Correction to be kept for hard labour for 10 days'.

John Rennie was asked to design a bridge of stone as stipulated in the Act. Unfortunately, the estimated cost of a stone bridge was more than the £300,000 allowed for in the Act. Consequently, a

The old Vauxhall Bridge of 1816 with Millbank Penitentiary on the north bank

new Act was passed in 1812 to enable the company to build an iron bridge. The company engineer was Sir Samuel Bentham, the brother of the philosopher Jeremy Bentham, and he started the design of an iron bridge. However, it seems there was a disagreement and Bentham was replaced by James Walker. Walker's bridge became the first iron bridge to cross the River Thames. It had taken over 30 years since the world's first iron bridge was constructed at Coalbrookdale for the new material of the Industrial Revolution to be used for bridging London's river. The bridge consisted of nine 78-foot cast-iron arches which stretched a total length of 809 feet, with a width of 36 feet. The bridge itself cost £175,000, but with all the approach roads and compensation payments the total cost was £297,000. At the opening ceremony on 25 July 1816, the chairman of the Vauxhall Bridge Company named the structure Regent's Bridge and expressed the hope that it would 'henceforth and for ever bear that designation'. This hope was not realised, and the bridge was henceforth known as Vauxhall Bridge because it was the most convenient crossing to Vauxhall Gardens.

James Walker (1781–1862)

James Walker was born in Falkirk, Scotland. After obtaining a degree at Glasgow University, he went to live in London, where he was apprenticed to his uncle, Ralph Walker, who was working on the engineering design of the East India Docks. On his uncle's death, he took over as chief engineer of the East India Dock Company. The design of Vauxhall Bridge was his first essay into bridge engineering. Following this, he won contracts for the maintenance of Westminster and Blackfriars bridges and was consulted by the several select committees set up to examine the question of whether to rebuild Westminster Bridge. When Thomas Telford died in 1834, Walker was elected the second president of the Institution of Civil Engineers. His other main achievements were in the design of lighthouses. As consulting engineer to Trinity House, he was responsible for building all the important English lighthouses of the time, including the famous Bishop Rock Lighthouse in the Scilly Isles.

The company was profitable despite having to pay compensation to so many interested parties and despite the construction of three other bridges, Lambeth, Chelsea and Albert, in the vicinity. By 1877, the annual income from tolls had risen from £4,977 in the first year to £62,392. The 1877 Metropolis Toll Bridges Act enabled the Metropolitan Board of Works to purchase all the privately owned bridges between Hammersmith and Waterloo in order to free them from tolls. The MBW paid £255,000 for Vauxhall Bridge, which was considerably less than its original cost. However, the bridge was no longer in good condition and soon afterwards the MBW discovered that the scour induced by the swift current through the narrow arches had exposed the timber cradles on which the pier foundations rested. Hundreds of cement bags were laid around the wooden cradles as a temporary measure to protect them. Not

147

surprisingly, when London County Council took over from the MBW in 1889, it was found that the cement bags were also being washed away and the piers were in danger of collapsing.

Since traffic across the bridge had nearly doubled since the MBW had taken it into public ownership, the LCC decided to build the replacement bridge much broader, with a roadway 80 feet wide. In 1895, an Act was passed enabling the estimated £484,000 to be raised from the rates of the whole LCC area rather than from local ratepayers alone, because the wider bridge was considered of benefit to the whole of London. Out of the total cost, £38,000 was required for the building of a temporary wooden bridge to allow traffic to cross the river while the new bridge was being constructed. The temporary bridge was opened in August 1898, allowing work to proceed on the demolition of the old bridge.

The new permanent bridge was designed by the LCC chief engineer, Maurice Fitzmaurice, with advice from the LCC chief architect, W.E. Riley. It consists of five steel arched spans with a total length of 759 feet. The plain design is enhanced by bronze figures of heroic size, which are located on both faces of each of the river-piers. These figures were designed by Alfred Drury and F.W. Pomeroy, who also designed the statue of Justice which stands on top of the Old Bailey. They represent Local Government, Education, Science, Fine Arts, Pottery, Engineering, Agriculture and Architecture. Unfortunately, the figures are little noticed because they are visible only from the river or the riverbank. On close inspection, the figures can be identified by what they hold. For instance, Agriculture clasps a scythe to her ample bosom, Science carries a primitive form of electric motor and Architecture holds a model of St Paul's Cathedral which is known by Thames watermen as 'Little St Paul's-upon-the-Water'.

The new bridge was not completed until 1906 because of a number of unforeseen difficulties. One problem arose during the construction of the south bank abutment when it was found that the River Effra, by now a storm relief sewer, flowed into the Thames at this point. The outflow was diverted to the north side of the bridge and today can be seen in front of Terry Farrell's

Vauxhall Bridge of 1906, showing the statue of Architecture

MI6 building. On the north bank, the River Tyburn, which is also a storm relief sewer, flows into the Thames about 200 yards upstream of the bridge, but this caused no problems. Seemingly, neither of these outflows causes much pollution, since the Thames Marine Mammal Survey records that a porpoise was spotted off Vauxhall Bridge in 2004.

The bridge was finally opened on 26 May 1906 by the chairman of the LCC, Mr Evan Spicer, to the accompaniment of music by the Battersea Borough Prize Band, which played popular favourites of the period, including excerpts from Wagner's *Tannhäuser*, Sullivan's *Pirates of Penzance* and Suppé's *Light Cavalry*, and ending with 'God Save the King'. Shortly after its opening, the bridge was crossed by trams on the double tramway in the middle of the carriageway. This was the first bridge to carry electrically powered trams across the Thames.

Once the bridge was completed, the temporary bridge could be demolished. The LCC advertised the sale of 40,000 cubic feet of good pinewood by tender for anyone willing to remove the whole of the structure which also contained 580 tons of scrap metal. Mr Charles Wall, who had won the contract for building the superstructure of the new bridge for £143,000, put in the highest bid at £50. Most other contenders offered no money or even demanded payment from the LCC. Therefore Mr Wall was awarded the contract to remove the now useless temporary bridge at a profit of £50 to the public purse.

During the Second World War, another temporary bridge was constructed about 200 yards downstream of Vauxhall Bridge. This was one of several temporary bridges built across the Thames in London because the Government feared that the Luftwaffe would target London's bridges and create transport havoc. The bridge looked precarious, with its steel girders supported by wooden stakes, but was in fact strong enough to carry tanks and guns. Fortunately, none of London's Thames bridges were severely damaged. Millbank Bridge was the first of the temporary bridges to be dismantled in 1948. Its steel girders were transported to Africa, where they were used to span a tributary of the Zambezi River.

Vauxhall Bridge has lasted 100 years with relatively few problems. However, no one could claim it is the most handsome of London's river crossings. In 1963, the Glass Age Development Committee commissioned a design for a more exciting structure, called the Crystal Span. This was to be a modern version of the old, inhabited London Bridge consisting of a seven-storey building enclosed in an air-conditioned glass envelope. The structure was to provide two three-lane carriageways for vehicles on the ground floor, with travelators for pedestrian access to shops, a hotel and a skating rink. There was also to be additional space for use by the nearby Tate Gallery, which was looking to expand its display area. This ambitious scheme was never executed, partly because of the estimated cost of seven million pounds.

The only major change to Vauxhall Bridge since its opening occurred in 1973, when the Greater London Council, which had recently taken over from the LCC, decided to introduce an extra traffic lane by reducing the width of the pavement. The GLC engineers needed to reduce the load on the bridge to balance

Vauxhall Bridge overlooked by the MI6 building

the increased traffic volumes. They proposed to achieve this by replacing the heavy, tall cast-iron balustrades with a lower and rather ugly box-girder structure. Lambeth and Westminster councils, which would not normally be expected to agree on anything, both objected. However, as reported in the *South London Press* of 22 June 1973, the GLC ignored their objections. Lambeth Council voiced its displeasure at what it saw as the high-handed approach of the GLC on the grounds that: 'Vauxhall Bridge is hardly in the top league as London bridges go and the parapet was its only redeeming feature. Our suggestion of a light-weight replica would have served the purpose just as well.' The GLC's parapet as seen today does not enhance the bridge's appearance and, despite being lower than the original one, still does not allow a view of the bronze statues standing at the face of the river-piers below.

Lambeth Bridge

In view of the great power of the archbishops of Canterbury, who in medieval times were second only to the monarch, it is perhaps surprising that until the 1860s there was no bridge between their palace on the south bank of the river at Lambeth and Westminster on the north bank. At Fulham, the bishops of London had helped initiate the building of a bridge as early as 1729, but it seems that the archbishops preferred their privacy and the income obtained from their ownership of the Lambeth horse ferry to the extra convenience of a bridge crossing to Westminster.

The first mention of the Lambeth horse ferry occurs in 1513 when the Archbishop granted the lease to Humphrey Trevilyan, with the proviso that the Archbishop, his officers, servants, goods and chattels should be transported free of charge. In contrast, the monarch was obliged to pay £100 a year for this privilege. The ferry was in a location so close to the seats of power of church and state that it was inevitably often used by some of the highest in the land, more than once with unhappy consequences.

In the reign of James I, many thought that Lady Arabella Stuart, the great-great-granddaughter of Henry VII, had at least as good a claim to the throne as did the King. In 1610, rumours spread that she was intending to marry William Seymour, who was the great-nephew of Lady Jane Grey, the successor to Edward VI as Queen of England, albeit for just nine days. James I was worried that this would enhance her claim to the throne and forbade the marriage. Nevertheless, they married in secret in July 1610. James was furious when he found out, and he ordered that she should be committed to the custody of Sir James Parry, who had a house in Lambeth. Next year, James decided to transfer her to the safer keeping of the Bishop of Durham, and it is recorded in the minutes of the Privy Council that he received her at Lambeth ferry and then took her to Durham in the north of England. She was later imprisoned in the Tower of London, where she died insane in 1615.

A tragic fate also awaited William Laud in 1633 when his party crossed the Thames via the horse ferry to Lambeth Palace on his appointment as Archbishop of Canterbury. The overladen boat capsized, throwing his servants and horses into the river. There was no loss of life, but Laud was a superstitious man and must have seen this accident as a bad omen. His fears were justified when in 1641 Parliament impeached him for blasphemy and forced Charles I to sign his death warrant. He was taken by river to the Tower of London, where he was imprisoned; shortly afterwards, he was executed on Tower Hill. In the ensuing Civil War, the Parliamentarians, under Oliver Cromwell, defeated the Royalists, and in 1649 Charles I himself was executed.

In 1656, Cromwell, who as leader of the victorious Parliamentarian forces had become Lord Protector of England, suffered a similar accident to that which had befallen Archbishop Laud. Cromwell's coach and horses were crossing the river on the ferry from Westminster to Lambeth Palace when the boat sank with his coach and horses, three of which drowned. Remembering what had happened to Laud, the public speculated as to whether this could be a warning of Cromwell's impending fall. Two years later, Cromwell died in Whitehall Palace and in 1660 the monarchy was restored in the person of Charles II, the son of Charles I.

Later, in 1688, the ferry was used by Mary of Modena to make a dramatic escape to France when her husband, James II, was about to be deposed. She left Whitehall Palace by coach with her baby son on the night of 9 December and drove to Millbank, where she crossed safely to Lambeth on the ferry. She stayed the night at an inn in Lambeth before driving to Gravesend and exile in France. James II himself left Whitehall two days later and threw the Great Seal of England into the river as he crossed over to Vauxhall, where a coach awaited him. Fortunately, a fisherman found the seal soon afterwards, by which time William and Mary were on the throne. James II's son survived and later tried to reclaim the throne in the unsuccessful invasion of 1715.

As we will see in the next chapter, from the late seventeenth century various attempts were made to obtain approval for a bridge at Westminster or Lambeth. Finally, in 1736, an Act was passed authorising the construction of a bridge at Westminster. This resulted in the closure of the Lambeth horse ferry and payment of compensation of £3,780 to the Archbishop and his lessees. From then on, vehicles crossed the river over Westminster Bridge, while foot passengers could cross by wherry from the stairs outside Lambeth Palace to the stairs in front of Market Street, today's Horseferry Road. Apart from the name of the road, nothing is left of the old horse ferry. However, the old ferry house survived well into the nineteenth century. Charles Dickens mentions it in *David Copperfield*, during the journey to Smith Square:

> There was, and is when I write, at the end of that low-lying street, a dilapidated little wooden building, probably an obsolete old ferry-house. Its position is just at that point where the street ceases, and the road begins to lie between a row of houses and the river.

The first serious proposal for a bridge at Lambeth was in 1809, but it was not successful, partly because preference was given to Vauxhall Bridge. In 1829, two proposals were put forward. One was for a stone bridge, to be called the Royal Clarence Bridge, and the other for an iron suspension bridge. The architect and civil engineer Charles Hollis argued strongly in favour of his proposal for a suspension bridge because iron was by then a well-tried material for bridge building and was considerably cheaper than stone. He suggested that 'the connecting of these two immediate neighbourhoods, and the facility which will be given for an intercourse between each other, as well as remoter parts of the town, must of necessity greatly increase the value of the property'.[20] The local historian Thomas Allen also expressed the hope that construction of the bridge would 'do away with several streets, filthy garrets and alleys, and various nuisances that exist in no part or neighbourhood of the metropolis in a greater degree than in this part of the parish of Lambeth'.[21] Although the Metropolis Suspension Bridge Act was passed in 1836, the project did not proceed.

In 1854, the Select Committee on Metropolitan Bridges recommended that four new bridges, including one at Lambeth, were necessary for 'providing further means of communication across the river'. The other recommended locations were at the Tower of London, St Paul's and Charing Cross. The men who put forward these ideas would be amazed that these schemes, which seemed essential in their day, took so long and were in two cases realised merely as footbridges. Tower Bridge was not completed until 1894; the bridge at St Paul's was never built as a vehicular crossing but finally appeared in 2001 as the Millennium Bridge; and the many proposals for a vehicular crossing at Charing Cross were never completed. Instead, the railway bridge at Charing Cross was enhanced by the Golden Jubilee Bridge in 2003. Lambeth Bridge was, however, approved and an Act was passed in 1861 'to build a bridge from Church Street, Lambeth, to Market Street on Millbank'. Market Street ran along the same line as today's Horseferry Road and the new bridge was to cross the river somewhat to the north of the former horse ferry. Finance was raised by the proprietors, who aimed to profit from the tolls. They had great hopes in this direction, since the £49,000 cost was much lower than that of other Thames bridges of the time.

Lambeth Bridge was designed by P.W. Barlow (1809–85), who had studied suspension bridges in America and Switzerland. Barlow's bridge used a system of diagonal bracing which was supposed to overcome the problems of instability often encountered with suspension bridges. He also adopted the American method of twisted wire cables instead of the eye-bar cables used in British bridges such as those at the Menai Straits and Hammersmith. The bridge underwent a stress test to demonstrate that it could handle weights of up to 800 tons, which was 6 times the maximum expected load. It was opened in November 1862 when a local entrepreneur, Mr Hodges, crossed in his new fire engine, followed by a mad rush of people, who caused much less vibration than at the opening of Brunel's Hungerford Bridge in 1845.

The great hopes for Lambeth Bridge were not realised. Tolls were much lower than expected because the road connections were not convenient for carriages and it was mainly used by pedestrians, who paid only one penny to cross. The bridge also

1866 view of the old Lambeth Bridge looking towards Millbank

deteriorated, and by 1877 when the MBW took it over the proprietors received only £35,974 compensation, which was considerably less than it had cost. Much rust was found in the twisted wire cables and in the wrought-iron girders. Even more worrying was the report on the state of the bridge produced by Sir Benjamin Baker, the engineer responsible for the recently completed Forth Bridge. He identified a significant tilt in the abutment tower on the Millbank side, which had resulted from the strain of the cables. Remedial work was undertaken, but in 1905 the LCC set a weight limit of 2.5 tons and a speed limit of 5 mph. In 1910, the bridge was closed to vehicles and all traffic had to be diverted to Westminster or Vauxhall bridges.

Nothing was done to the emasculated Lambeth Bridge until the LCC Improvements Committee produced a report in 1923 in which London was compared unfavourably with Paris with regard to its river crossings. With Lambeth Bridge out of action, London had just three river crossings between Blackfriars and Vauxhall – a distance of about two miles. In Paris, there were twelve bridges over the River Seine covering a similar distance.

An Act of Parliament was passed the following year authorising the demolition of the old suspension bridge and the construction of a new bridge, on a line slightly further upstream in order not to impinge on the ancient Lambeth Palace and St Mary's Church on the Lambeth side of the river. The contract was signed and work started in 1929, by which time two severe neo-classical office blocks had risen on the Westminster embankment on either side of Horseferry Road. They were London's largest at the time. The northern block was built for the new conglomerate ICI, and it is no longer a secret that the southern block is now occupied by the security service, MI5.

Considerable effort was put into the design of the bridge, which was to provide a link between the seat of the head of the Church at Lambeth and the powers of the state and big business on the Westminster side of the river. Sir George W. Humphreys, the LCC chief engineer, was responsible for the engineering design, and he collaborated closely with Mr G. Topham Forrest, the LCC chief architect, and with the eminent architect Sir

The Houses of Parliament viewed through the arches of Lambeth Bridge of 1932

Reginald Blomfield. The first design, which was strongly supported by Blomfield, was for a reinforced concrete bridge with elliptical arches faced with granite. However, a large-scale model constructed on wasteland near by indicated that this design would be difficult to realise in view of the gradient required to provide sufficient headroom for passing river traffic. The final design was for a steel bridge of five arches carrying a 60-foot-wide roadway. Blomfield's main contributions were to disguise the steel skeleton of the bridge arches by covering them with flat steel panels, and to install obelisks topped with carved pineapples on either side of the approaches.

There was much speculation about the significance of the pineapples. Some thought that they might refer to the famous seventeenth-century gardener John Tradescant, who had a house in Lambeth and today is commemorated in the Museum of Garden History which occupies the former church of St Mary at Lambeth. However, there is no evidence that pineapples were grown in this country until the eighteenth century. The *South London Press* decided to investigate, and in the issue of 15 July 1932 a spokesman for Sir Reginald stated that 'they have no name, they do not symbolise anything and they are put there for no reason except to decorate the bridge'. According to an article in *Country Life* of 18 February 1965, this sort of pineapple finial is quite common, but it really resembles a pine cone more than a true pineapple, which has saw-edge leaves branching out from its apex. Similar pine cone finials have been found on Roman sites, as they were a fertility symbol of the rather bloodthirsty and warlike cult of Mithras. Today, it is generally accepted that pineapples are a sign of welcome arising from the Jamaican custom of placing a pineapple on a house as a symbol of friendship to travellers.

The total cost for building the bridge and its approaches amounted to £936,365, of which the bridge itself cost £555,029. Finance was provided by the LCC with a 50 per cent grant from the MOT. The contract was awarded to Dorman, Long & Co. Ltd in December 1928. The project was completed in July 1932. On 19 July, the new bridge was opened by George V and Queen Mary to the accompaniment of the bands of the Life Guards and

Pineapple obelisk at the approach to Lambeth Bridge

Grenadier Guards, as massed crowds lined the nearby streets and the pavements of the bridge.

Twenty years later, another royal occasion gave rise to a very different scene. George VI, whose reign spanned the Second World War, had just died and lay in state in Westminster Hall. Thousands queued silently in the bitter February weather to pay homage to the King, who had been almost as much a figure of hope to the British people during the war years as Sir Winston Churchill. The line stretched for over four miles, from Westminster, along Millbank, across Lambeth Bridge and along the Albert Embankment on the other side of the river. It is said that his daughters, the present Queen and the late Princess Margaret, stood for several minutes in Westminster Hall mourning in the background while members of the public paid their last respects to the dead king.

By the 1990s, concerns were being expressed about the state of Lambeth Bridge. Westminster Borough Council, which is today responsible for maintenance, asked Kumar Associates to do a load-capacity assessment and to design any necessary strengthening works. In his paper to the Institution of Civil Engineers, recorded in their *Proceedings* for May 2003, Mr A. Kumar stated that Lambeth Bridge represents one of the finest examples of workmanship in the country. However, the survey had identified one design problem, as well as corrosion resulting from continual tidal submersion of the steel arches near the abutments. It was agreed that the necessary repairs should be carried out and the bridge repainted without any disruption to traffic.

The project lasted two years and cost two and a half million pounds. All corroded parts were replaced from underneath the bridge, using temporary braced frames for support. Repainting proved difficult because it was found that up to 19 layers of paint had bonded too strongly to the steel frame for them to be removed without great expense. Therefore it was decided to remove loose paint by grit-blasting and apply five coats of modern spray paint to match the original colours as far as possible.

The red colour scheme of Lambeth Bridge contrasts with the light-green colour scheme of Westminster Bridge. The reason for this arises from the internal layout of the Houses of Parliament. The House of Lords, with its predominantly red decor, is located to the south of the building, near Lambeth Bridge, while the House of Commons, with its green benches, is located to the north, near Westminster Bridge. Just as the Commons has much more power than the Lords, so it must be admitted that Westminster Bridge is far more important than Lambeth Bridge as a Thames crossing. However, Lambeth Bridge, with its historical association with the ancient horse ferry and Archbishop's palace, provides an elegant viewpoint for one of London's greatest sights – the Houses of Parliament.

CHAPTER 8

Westminster

The present Westminster Bridge, with its seven iron spans supported by granite piers, was opened in 1862. It replaced an earlier stone bridge of 1750, which had been the first bridge to be built over the river in central London since the construction of Old London Bridge in the thirteenth century.

For a thousand years, Westminster has been the centre of government, at first of England and then of the whole of the United Kingdom. A more unlikely place for such a centre of power would have been hard to imagine when in the eighth-century Benedictine monks first set up an abbey there. A purported charter of King Offa of Mercia granted the monks lands on the 'isle of Thorney in the terrible place that is called Westminster'. The lands were a desolate swamp surrounded by two forks of the River Tyburn, which used to flow into the Thames here from its source in the hills of Hampstead. One fork entered the Thames to the south of the abbey gardens by today's Great College Street, and the other flowed along an almost direct line from today's St James's Park to the site of Westminster Bridge. The River Tyburn has long since disappeared in Westminster, but another branch of this river which flowed further west through Pimlico still exists as an overflow sewer.

In the eleventh century, Westminster was chosen by Edward

the Confessor as the location for his royal palace and a massive new abbey church, images of which can be seen in the Bayeux Tapestry. From the time of the Norman Conquest in 1066, all royal coronations have taken place in Westminster Abbey, and Parliament has sat at Westminster since it was first established by Edward I in 1295. Even today, the Houses of Parliament are called 'the Palace of Westminster' in recognition of their original function as a royal palace. In the sixteenth century, Henry VIII moved his palace from Westminster to Whitehall and when Whitehall Palace was destroyed by fire in 1698 the area became the site of many new government offices, including the home of the first prime minister, Robert Walpole, in Downing Street.

By the eighteenth century, London stretched along the banks of the Thames as far as Horseferry Road, and new estates were developed further north in Piccadilly, St James and Bond Street by wealthy aristocrats such as the Earl of St Albans and Sir Thomas Bond. Many great aristocratic mansions, including Burlington House (later converted into the Royal Academy), Devonshire House and Berkeley House, were located in these areas, but the vicinity of the Palace of Westminster was also popular with politically minded aristocrats. The Earl of Pembroke, who like Lord Burlington was an amateur architect, had a house constructed there in 1724 in the Palladian manner to the design of Colen Campbell. Pembroke later became heavily involved in the promotion and execution of the project to build Westminster Bridge.

The story of a river crossing at Westminster begins in Roman times. At the time of Caesar's invasion of Britain in 55 BC, the tide did not reach as far as Westminster. The river was wider and the water level much lower than today, and this is one of the sites which might have been used by Caesar to cross over on his way to attack the Iceni, the important British tribe based in East Anglia. Once Roman rule was established after Claudius's invasion in AD 44, Watling Street, the main road from the south, may have run along the south bank of the river to a ford at Westminster, on its way up the Edgware Road and on to Chester.

Archaeologists have long disputed this matter, and even if there was a ford here in Roman times, it did not last long. There is a medieval legend that the first church at Westminster was consecrated by St Peter himself in the seventh century. He arrived incognito on the south bank and asked a fisherman named Edricus to ferry him across the swollen river for the ceremony. On the return journey he commanded Edricus to cast his nets into the Thames and they were immediately filled with a vast quantity of salmon, proving St Peter's identity. Until 1382, the event was celebrated by an annual offering of salmon at the high altar on the anniversary of the Abbey's consecration. The figures of salmon inlaid into the tiles on the Chapter House floor are also believed to commemorate this occasion.

It seems incredible that no permanent river crossing was built at Westminster until 1750. Until then, Old London Bridge was the only permanent river crossing in the central London area. Until the construction of Fulham Bridge in 1729, travellers had to go all the way to Kingston to avoid the horrendous congestion on London Bridge. By 1700, London had grown into the largest city in the world, with a population of 600,000, many of whom lived in the area of today's City of Westminster. Having crossed London Bridge with difficulty, the traveller was confronted by daunting traffic jams along Fleet Street and through Temple Bar on the route to Westminster, as well as uneven and dangerous road surfaces. Lord Tyrconnel, commenting on the state of the road from the City of London in a House of Lords debate, declared that 'the passenger is everywhere surprised or endangered by unexpected chasms, or offended or obstructed by mountains of filth'.[22]

The first serious attempt to obtain authority for the construction of a bridge at Westminster was made in 1664, when Charles II presided over a meeting of the Privy Council to discuss the matter. Strong reasons were put forward in favour of a new bridge, quite apart from its obvious convenience in terms of the lack of alternative crossings in the area: it would benefit the King in his travels between his several riverside palaces between Hampton Court and Greenwich; it would bring extra trade to Westminster, which, despite the presence of the court, had a large number of poor inhabitants; and it would provide an alternative crossing for troops in case of a possible rebellion in the unruly district of Southwark. The arguments against the bridge were put forward by the City Corporation, the watermen and other vested interests. They claimed that London Bridge was perfectly adequate but that if a new bridge were built at Westminster, London Bridge would fall into decay through loss of toll revenue, and the inhabitants of Southwark would be thrown into poverty. The water level in the river would rise and cause flooding along the riverbanks from Whitehall to Chelsea, and many watermen would lose their jobs, thus depriving the navy of a ready supply of sailors in time of war.

None of these suppositions was credible, but the opponents of the bridge came up with one argument that finally clinched the matter, in the form of an unsecured and interest-free loan of £100,000 to the King from the City Corporation for his help. They also offered

> most humble thanks for the great Instance of His Majesty's goodness and favour towards them as expressed in preventing of the new Bridge proposed to be built over the River of Thames betwixt Lambeth and Westminster which is considered would have been of dangerous Consequence to the State of the City.[23]

Charles II accepted the flattery and the bribe with grace, and the proposal was rejected, although soon afterwards the City did introduce a rule to improve traffic flow on London Bridge, insisting that all vehicles should keep to the left.

No further serious proposals for building a bridge at Westminster were put forward until 1721, but again opponents managed to defeat the Bill in Parliament. However, in 1734 a new group called a 'Society of Gentlemen' was formed with the aim of promoting the new bridge, for which Nicholas Hawksmoor (1661–1736) produced a design. Hawksmoor was at the time working on the design of the two western towers of

Hogenberg's sixteenth-century map of London with the single river crossing at London Bridge. A second central London crossing, at Westminster, was not constructed until 1750, despite London's massive expansion westwards during the following 200 years.

Westminster Abbey, which were completed in 1745, and it was his design that was initially put forward when a Bill was finally presented on 23 February 1736. With the growing prosperity of Georgian London and the expansion in foreign travel, the reactionary attitudes of the bridge's opponents were increasingly seen to be unreasonable and antagonistic to progress. Patriotic fervour was aroused by the knowledge that London was far behind other European capital cities, most of which had several bridges while London had just one.

For a brief moment, one of the points raised against the bridge, that it would cause flooding along the riverbanks, did seem to have some merit. During the debate on the Bill, the river did overflow its banks, and Westminster Hall, where judges were sitting in court in their ermine robes, was inundated with water two feet deep. The judges had to be carried out to safety. Westminster Hall had endured flooding on a number of occasions in the past, but the 1736 flood was the last one recorded and despite the indignities suffered by the judges it did not affect the outcome of the debates. The Act for 'Building a Bridge across the River Thames from the New Palace Yard in the City of Westminster to the opposite Shore in the County of Surrey' received Royal Assent on 20 May 1736.

The Act appointed a distinguished body of commissioners, among them the Archbishop of Canterbury and the Lord Chancellor, with powers to decide on the design of the bridge and what materials were to be used, as well as to purchase land for building approach roads and to pull down houses as necessary. As was usual with the early bridges, punishments were laid down for anyone who attempted to destroy the bridge or endanger the lives of passengers. In this case 'they should be treated as felons and suffer death without benefit of clergy'. Compensation was to be paid to the Archbishop and his lessees for the loss of the Lambeth horse ferry, and to the watermen who ran a Sunday ferry across the river at Westminster. The Sunday ferry crossed from stairs by New Palace Yard which are marked on early maps with the confusing name of 'Westminster Bridge'. It seems that landing stages that had sizeable jetties built out into the river were called 'bridges' rather than the normal name, 'stairs'.

The most intriguing part of the Act was devoted to the method of finance. Since the bridge was promoted as of national importance and not just for the convenience of the inhabitants of Westminster, it was out of the question to raise the finance by local taxation. The usual method of involving private enterprise and charging tolls was also rejected in favour of a lottery. Lotteries were in vogue at the time and had been used for various enterprises including fighting wars. However, lotteries were often subject to abuse and fraud, and some considered them immoral and a danger to society. Many people had suffered huge losses from the collapse of the South Sea Company, which became known as the South Sea Bubble, and although this involved investing in a business enterprise, such an investment seemed much more like a lottery. William Hogarth certainly thought so, and two of his earliest engravings, entitled *The South Sea Bubble* and *The Lottery*, mock the greed and chicanery involved.

The first Westminster Bridge lottery was a failure, and further Acts of Parliament were required to set up four additional annual lotteries. One most unusual prize was offered in the 1738 lottery, with the aim of encouraging more people to buy tickets. This was the largest-ever silver wine cooler, designed by the famous sculptor John Michael Rysbrack. The winner sold it immediately to the Hermitage Museum in St Petersburg, but before it went out of the country a silver-plated copy was made and this can be seen today in the Victoria and Albert Museum Silver Galleries. After the fifth lottery, additional finance was provided by government grants. The final cost of the whole project has been estimated at £389,500, which was considerably more than any other eighteenth-century Thames bridge. The use of the lottery to finance Westminster Bridge prompted Sir Henry Fielding to call it 'the bridge of fools' and this name stuck when the project later dragged on for much longer than was forecast.

Decisions about how to design and build a bridge in such an important location took up much of the commissioners' time and caused considerable controversy. Although British architects had shown themselves able to handle great buildings, especially after Sir Christopher Wren's masterpiece, St Paul's Cathedral, there was no body of expertise in the construction of bridges over major rivers. It is true that Old London Bridge, built over 600 years before, had been considered one of the wonders of the world, but it was by now hopelessly old-fashioned. The wooden bridge at Fulham was not considered a sufficiently imposing model for Westminster either. The situation in France was more advanced. Specialist bridge engineers, including the great Jean Perronet, who would design the Pont de Neuilly over the Seine in Paris, were establishing standards for bridge construction and disseminating these through the Département des Ponts et Chausées. The Westminster Bridge commissioners must have been tempted to request French expertise, but in the end it was the Swiss Charles Labelye who in 1738 was appointed chief engineer.

Charles Labelye (1705–81)

Charles Labelye was born in Vevey in Switzerland, although his father was French. In 1721, he came to England without one word of English, as he wrote in 1751 in his *A Description of Westminster Bridge*. He became heavily involved with Freemasonry and presumably used his time in gathering building experience as well as being a leading member of the French Masonic Lodge. Little is known of his career during the years up to his involvement with the Westminster Bridge project. He did some work on harbour design and on drainage of the Fens. His only successful bridge design had been for a small stone bridge over the River Brent. In 1736, he produced one of several plans to improve navigation through London Bridge. Labelye's proposal was to halve the number of piers by removing every alternate arch, thus doubling the free waterway under the bridge. However, none of these plans was implemented.

Westminster Bridge was Labelye's greatest achievement, which he liked to compare with Wren's St Paul's Cathedral as a triumph of British architecture. Labelye fell ill soon after the bridge was opened, possibly from the tremendous strain of the 12-year construction project, and moved to the south of France to improve his health. He was never to return to England and died in Paris in 1781.

It is something of a mystery why the relatively inexperienced Labelye was chosen as chief engineer for so important a project, but at the time he was no less qualified than the other possible contenders. To judge from Labelye's writings about the Westminster Bridge project, he had boundless confidence in the correctness of his own engineering decisions, and the only sign of modesty he showed was in his apology for his poor English. What seems to have impressed the commissioners was his detailed calculations about the effect of the tide on the stone piers of Hawksmoor's Westminster Bridge design. The calculations showed that with only eight river-piers, as compared with the nineteen piers of London Bridge, the tide would hardly create any disturbance for navigation or cause much scouring of the foundations.

Labelye was also involved in discussions about exactly where the new bridge should be built and what materials should be used. Although the 1736 Act had stipulated that the bridge should cross the river from New Palace Yard, which is about 50 yards south of today's Westminster Bridge, some, among them Charles Labelye, still preferred a crossing at the site of the Lambeth horse ferry, where the river was 300 feet narrower. Others wanted the bridge to be sited further downstream, at Whitehall or even Charing Cross. Since New Palace Yard was most convenient for members of both houses of Parliament, it was always likely that this site would be chosen, but for some reason it was eventually decided to build the bridge 50 yards downstream at the Woolstaple. There had been a wool market there from medieval times, and although this was discontinued in the fourteenth century, the name was preserved until the construction of Westminster Bridge.

Having finally agreed the location, the commissioners had to decide on the design of the bridge. Several proposals for bridges of wood, stone or a combination of the two were submitted. Many of the commissioners favoured a wooden bridge because of the far lower cost. This resulted in much protest in the press and, as Labelye wrote later, 'the public were highly disgusted at the thoughts of having a Wooden Bridge in the Metropolis of the British Empire'. Nicholas Hawksmoor put it more pithily: 'We are striving to have it in stone, but there are some wooden-headed fellows endeavouring to have a wooden one.' In 1738, the initial decision was taken to build the piers and abutments of stone, but for the superstructure to be constructed of wood. The Earl of Pembroke, who by now was one of the most influential of the commissioners, insisted that the stone piers should be built in such a way that they could support a stone superstructure in case it was later decided that this was preferable, and it was he who later laid the first stone, on 29 January 1739.

The contract for the wooden superstructure was awarded to James King and his partner John Barnard, who later constructed the first Kew Bridge. It must have seemed that Westminster was to have a cheap wooden bridge after all, but King and Barnard were never given the chance to go ahead. A major reason for the abandonment of the wooden bridge design was the damage caused by the severe winter of 1739–40 to the wooden piles which had been driven into the river-bed to support the construction of the two centre piers. The Thames froze solid from Boxing Day to the middle of February, and, as on previous such occasions, Londoners took to the ice to set up a variety of entertainments for a Frost Fair. One popular activity was to walk over the river to the recently constructed centre piers of Westminster Bridge, where ladders allowed sightseers to climb on top. It is unthinkable today that the public would be permitted access to a partially completed building site, but it seems the sightseers came to no harm. When the ice finally melted, it was found that all 140 wooden piles had been broken or carried away by ice floes. This confirmed the opinion that wood was not the right material. Labelye produced his design for a stone bridge of 13 semicircular arches. This was accepted by the commissioners, who now gave him responsibility for the superstructure as well as for the foundations.

The commissioners paid compensation to King and Barnard for the loss of the construction contract and appointed Andrews Jelfe and Samuel Tuffnell as contractors for the stonework, under Labelye's supervision. The most innovative and, as it turned out, controversial decision made by Labelye was to build the river-piers within reusable wooden caissons rather than in the more traditional cofferdams. The caissons were enormous wooden boxes which were constructed on the side of the river and floated out to the location of a pier. After the river-bed had been dredged to the shape of the pier foundations, the caisson

170

was sunk into the cavity and building work could then be carried out in the dry. Once the level of the pier reached above the water level, the sides of the caisson were detached from the bottom and raised up so that they could be reused for another pier. Caissons had been used before, and Labelye admitted he did not invent them, but no one had built caissons so large or had thought of the technique of reusing them. However, Batty Langley, one of Labelye's competitors for the design contract, did claim with some justification that he had thought of this idea first in his proposal of 1736. Batty Langley was a prolific eighteenth-century architect-builder who produced pattern books for the design of houses which ordinary builders could use in the vast number of housing developments undertaken at the time, especially in London. He was not pleased to have lost out to an upstart foreigner and complained bitterly that Labelye had stolen his idea for the caissons. He later used every opportunity to attack Labelye's methods.

Unfortunately, Labelye did make a critical error in his evaluation of the river-bed, which he thought consisted of a firm layer of gravel but which turned out to be clay. Because of this, he decided it was not necessary to drive piles into the river-bed to support the foundations and just sank his caissons a few feet into the dredged clay bottom. The Earl of Pembroke laid the last stone in October 1746, but the very next year one of the piers showed signs of settlement. When a massive stone block fell down from one of the arches supported by this pier, it seemed that the whole bridge might collapse. Labelye had been criticised by a number of people, including Batty Langley, for not taking enough care with the pier foundations. He had attacked these critics with vigour in a letter of 1743 for spreading rumours that were 'fake, malicious and scandalous', so it is not surprising that his critics hit back with equal vigour when they were proved to have been right. The fiercest invective came from Batty Langley in a pamphlet entitled *A Survey of Westminster Bridge as 'tis now sinking into ruin*. Langley lived in a house by Parliament Stairs and had a good view of the construction of Westminster Bridge by his rival. He must have been overjoyed at the sight of the sinking pier, but many thought his criticism went too far when he wrote about Labelye: 'It would have been happy for the public had you been hanged before Westminster Bridge was thought on.'

The commissioners were clearly embarrassed by the sinking pier, but the Earl of Pembroke helped ensure that Labelye kept his job, and eventually the pier and two affected arches were removed and rebuilt with stronger foundations. Labelye, always strong on self-justification, pointed out that it was possible to remove just the two arches without affecting the rest of the bridge because of his forethought in ensuring that each arch could stand up independently without needing the lateral support of an adjacent arch, as would have been the case if the arches were elliptical rather than semicircular. Labelye's reputation received an unexpected boost when in 1749 two earthquakes caused havoc in Westminster but left his bridge standing despite the incomplete work on the failing pier, 'to the great Amazement of many, and the no less Confusion and Disappointment of not a few malicious and ignorant People, who had confidently asserted that upon unkeying any one of the Arches the whole Bridge would fall'.[24]

The Westminster Bridge project had dragged on much longer than had originally been forecast, but Labelye pointed out that there were many special circumstances. From 1740 to 1748, England had been at war with France and Spain in the War of the Austrian Succession. This war was ostensibly about Maria Theresa's claim to the Austrian throne but really concerned which European power was to be dominant on the world stage. The impact of the war on Westminster Bridge was considerable, because the vast amount of Portland stone required had to be transported by sea from Dorset. Labelye claimed that since Westminster Bridge was constructed of solid stone rather than the normal mixture of rubble faced with stone, the quantity of stone needed was nearly double that used for St Paul's Cathedral. Press-gangs were on the lookout for healthy seamen to enrol in the navy, and Labelye's men were reluctant to risk the long voyage along the Channel and up the Thames without escort, thus delaying the supply of stone.

Another, potentially even more dangerous situation arose with the Jacobite invasion of 1745. The leader of the Jacobites was Bonnie Prince Charlie, known as the Young Pretender. His grandparents, James II and Mary of Modena, had escaped from Westminster to France via the Lambeth horse ferry after being deposed by William and Mary, as described in Chapter 7. The Jacobite army consisted mainly of Scottish Highlanders, who met little resistance as they swiftly advanced as far south as Derby, inducing panic in London. However, the Jacobites found little support in England for their cause of replacing the Protestant George II with the Roman Catholic Bonnie Prince Charlie. English patriotism was aroused by the newly composed anthem, 'God Save the King', and the Jacobite army had to retreat northwards and were soon defeated with great cruelty by George II's son, the Duke of Cumberland, at the Battle of Culloden. The effect of the Jacobite threat on Westminster Bridge therefore turned out to be minimal.

Apart from wars, earthquakes and the disaster with the sinking pier, there were many accidents and even acts of sabotage which caused further delays. It was not uncommon for large and unwieldy barges, carrying cargoes of 100 tons or more, to lose control and crash into the construction works. Some of these accidents were undoubtedly deliberately caused by disgruntled watermen who had always opposed Westminster Bridge and would do anything to abort or delay a project which they considered, with some justification, was injurious to their livelihood. One such act of sabotage occurred in 1739 when someone pulled out the plug of the barge which was being used to dredge the river-bed in preparation for laying the caissons, causing the vessel to sink. The designated punishment of death, as laid down in the Act of 1736, does not seem to have been a deterrent, probably because offenders did not expect to be caught by the ineffective watchmen. There was no official police force in London until 1829 when Robert Peel, then Home Secretary, set up the Metropolitan Police. In the eighteenth century, policing was done mainly by elderly or infirm men who could not obtain other paid employment and who were mocked in a famous cartoon by Thomas Rowlandson and given the nickname of 'Charleys'. The saboteurs of 1739 therefore seemingly went undetected and unpunished.

Injuries to workers and passers-by were not uncommon. The local Westminster Infirmary, where most of the injured people were treated, claimed that they handled up to three patients a day, including some fatalities. The commissioners normally paid compensation in the case of disabling injuries or death.

Westminster Bridge was finally opened on Sunday, 18 November 1750, nearly 12 years after the laying of the first stone. Sadly, the Earl of Pembroke, who had been such a strong supporter of the bridge and of its designer, had died just before. It was strange that the commissioners chose a Sunday to open the bridge, since many held strong views on the strict observance of the Sabbath. Consequently, it was decided to conduct the ceremony at midnight on the Saturday. However, since the ceremony lasted a full two hours after many attendees had participated in a convivial dinner at the nearby Bear Inn, this particular Sunday went off with a bang as the new national anthem was sung to loud applause.

Comments on the new bridge were mostly favourable. Labelye's own assessment was typically adulatory:

> The Bridge has certainly nothing of the Kind in Europe, and perhaps in the whole world, that can be brought into Competition, and much less exceed it . . . The Bridge will want no considerable Repairs for a long Course of Years.[25]

Although the prediction regarding its longevity proved over-optimistic, Old Westminster Bridge was a beautiful structure, as can be seen from the numerous paintings by Canaletto, Samuel Scott, Antonio Jolli and others. One of the most dramatic views of the bridge was painted much later by J.M.W. Turner, who watched the catastrophic fire which burnt down the old Palace of Westminster in 1834. Turner's viewpoint is on the Lambeth side of the Thames, from where he captures both the raging inferno that engulfed the palace buildings in the background

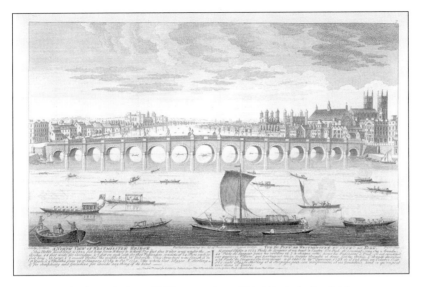

Charles Labelye's 1750 engraving of his Westminster Bridge

and the massive structure of Westminster Bridge defiantly looming out of the fire to cross the river in the foreground. Turner exaggerates the slope of the bridge, which is impossibly steep in the painting, but it is true that there were many complaints about the difficulty of crossing the bridge in horse-drawn vehicles, especially in wet weather, because of its sharp incline.

The most famous literary reference to Westminster Bridge is contained in the title of William Wordsworth's famous sonnet 'Composed upon Westminster Bridge', although the actual poem is about the view of London and the Thames, and does not mention the bridge itself:

> Earth has not anything to show more fair:
> Dull would he be of soul who could pass by
> A sight so touching in its majesty:
> This City now doth, like a garment, wear
> The beauty of the morning; silent, bare,
> Ships, towers, domes, theatres, and temples lie
> Open to the fields, and to the sky;

175

All bright and glittering in the smokeless air.
Never did sun more beautifully steep
In his first splendour valley, rock, or hill;
Ne'er saw I, never felt, a calm so deep!
The river glideth at his own sweet will:
Dear God! The very houses seem asleep;
And all that mighty heart is lying still!

It is perhaps surprising that this patriotic eulogy to England's capital city should have been written by Wordsworth, the great poet of nature and Romanticism, and one-time supporter of the French Revolution. However, the sonnet was written in 1802, by which time Wordsworth, like many other English radicals, had become disillusioned with the Revolution. It is not so likely that Wordsworth would have been inspired to write these lines later in the nineteenth century, when Claude Monet painted his famous view of the new Westminster Bridge through a variegated filter of fog, which had replaced Wordsworth's 'smokeless air'.

Although so near to the centre of government, Westminster Bridge had an unsavoury reputation for crime in the eighteenth and nineteenth centuries. This was partly because of the domed octagonal recesses which Labelye had thoughtfully installed over each of the piers to provide shelter for pedestrians in wet weather. Unfortunately, they also provided a hiding place for robbers, who, as already mentioned, were unlikely to be caught by the elderly and frail watchmen. As well as the problems with robberies, numerous cases of suicide occurred. One near-suicide was recorded in the magazine *City Limits* of 11 February 1888. Casanova had come to London in 1763, but it seems that for once his charms had let him down, and he became depressed. He was close to bankruptcy and was threatened with incarceration in Newgate Prison. Moreover, he hated London, where inns served bitter beer rather than wine and people pissed in the streets. He went to Westminster Bridge, having loaded his pockets with weights, and was gazing mournfully into the Thames when the well-known fop Sir Wellbone Agar, returning over the bridge from a night out in the red-light

ABOVE: Richmond Bridge viewed from the south

BELOW: Richmond Footbridge, Lock and Weir

ABOVE: Chiswick Bridge

BELOW: Tierney Clark's Marlow Bridge, modelled on the old Hammersmith Bridge

ABOVE: Ornamental cast-iron shields on a Battersea Bridge pier

BELOW: Lambeth Palace from Lambeth Bridge

RIGHT: One of Chelsea
Bridge's galleon
lamp-posts

BELOW: Albert Bridge

LEFT: Lamp-post and view from Westminster Bridge

BELOW: Embankment Place and Hungerford Bridge

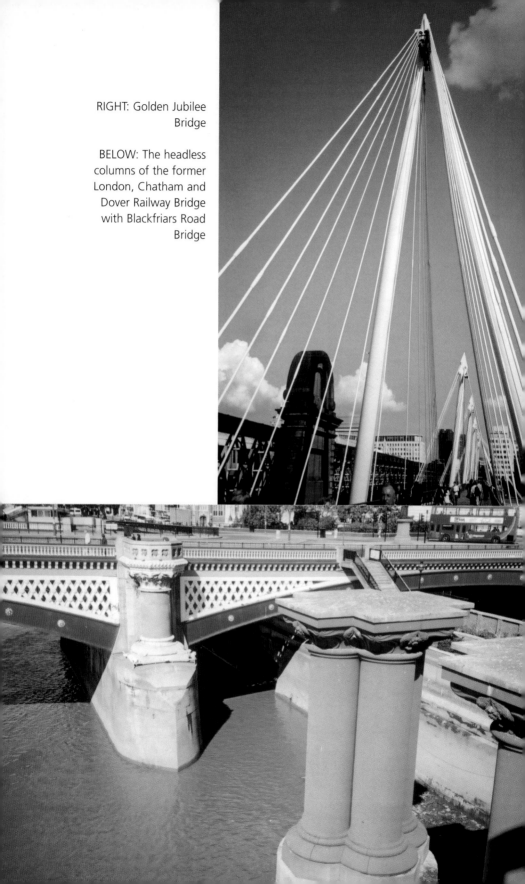

RIGHT: Golden Jubilee Bridge

BELOW: The headless columns of the former London, Chatham and Dover Railway Bridge with Blackfriars Road Bridge

ABOVE: Insignia of the London, Chatham and Dover Railway by the southern end of Blackfriars Bridge

LEFT: Millennium Bridge

ABOVE: George III's coat of arms, salvaged from Old London Bridge

BELOW: The massive suspension chains of Tower Bridge framing the Tower of London and the 'Gherkin'

district in Southwark, persuaded Casanova to accompany him to a strip show at an inn near Trafalgar Square. This banished all further thoughts of suicide, and Casanova left England surreptitiously soon afterwards to restart his amorous career on the Continent.

Concerns were raised about the safety of Labelye's beautiful bridge as early as 1759, when the two central arches of Old London Bridge were replaced by a wider single arch, thus increasing the flow of the tide towards Westminster. Later reports commissioned from the eminent engineers John Rennie, Thomas Telford, Isambard Kingdom Brunel and James Walker all confirmed that the foundations were being eroded, especially after the effect of the tide was magnified by the removal of Old London Bridge in 1832. No fewer than seven select committees were set up to examine the state of Westminster Bridge and to decide what action to take.

The conclusion of the 1844 committee was that 'on a review of the Evidence, no case has been made out to justify the Committee in recommending to the House the pulling down of the present Bridge, and the constructing of a new one'. However, the chairman, Sir R.H. Inglis, stated his opinion that the opposite conclusion should have been drawn, since all the expert witnesses had in fact concurred that 'the foundations of Westminster Bridge having been originally vicious, the bridge can never be sound'. The majority decision was to continue propping up the old bridge rather than incur the expense of constructing a new one. When, eventually, in 1853 a new select committee finally decided that it was essential to rebuild Westminster Bridge, the total cost of repairs to the old bridge between 1810 and 1853 had amounted to £200,000, which was about half the cost of reconstruction.

Labelye's bridge had been built when Britain was vying with France and Spain for domination in Europe and the acquisition of colonies in America, Africa and Asia. The predominant architectural style was classical and Westminster Bridge, with its 13 semicircular arches of Portland stone, was eminently classical in design. Since that time, despite the loss of the American colonies, Britain had grown into the greatest international power, with an empire that covered a quarter of the globe. When the ramshackle Palace of Westminster was burnt down in 1834, Charles Barry and Augustus Pugin won the commission to rebuild it in the newly fashionable Gothic style, which was seen to represent the true spirit and glories of Elizabethan England, as opposed to the classical style, which was viewed as a foreign import.

Barry (1795–1860) cooperated with Thomas Page, the architect of Chelsea Bridge, in the design of the new Westminster Bridge to ensure that it accorded with the new Houses of Parliament. Page's Westminster Bridge has seven elliptical iron arches supported by piers consisting of massive 30-ton granite blocks. Barry inserted Gothic quatrefoils in the spandrels of the arches and attached ornamental shields emblazoned with the arms of England and Westminster. The central arch has a span of 120 feet, and the side arches reduce in width to 94 feet at the abutments. The use of elliptical arches gave the bridge a much flatter gradient than Labelye could achieve with his semicircular arches. Vehicles benefited not only from the gentler slope, but also from the vastly increased width of the roadway, which provided an unprecedented 55-foot carriageway and two 15-foot footpaths. Page even installed a kerb in the centre of the carriageway to ensure that the two lines of traffic would not interfere with each other, as happened in so many London streets at the time despite the 'keep left' rule which had first been introduced on London Bridge in 1722. The kerb was later replaced by a central tramway, which was removed in the 1930s to leave the widest carriageway of any of London's remaining Victorian bridges.

The construction project suffered considerable delays, partly due to the bankruptcy of the initial contractor, Messrs Mare & Co. Thomas Page then took over the contract himself and received much praise for completing the work without any significant interruption of traffic. This he achieved by removing only one half of the old bridge in order to construct the first part of the new structure while allowing traffic to cross via the

Westminster Bridge of 1862 with Big Ben

remaining half of the old one. Most of the vast quantity of Portland stone from the old bridge was sold off, but Page used some of it to construct the abutments. The new bridge was opened in 1862 on 24 May, which was Queen Victoria's birthday. The Queen had originally agreed to perform the opening ceremony herself, but went into prolonged mourning after the death of Prince Albert and so the rather muted celebrations went ahead without her.

The 1962 centenary was equally lacklustre. As reported in *The Times* of 25 May 1962, the only celebrations consisted of a lunch at County Hall, presided over by the chairman of the LCC, and the application of a new coat of battleship-grey paint, which was the standard colour of all of London's Thames bridges in the post-war period. In 1970, the bridge was repainted in the present green colour to match the colour scheme of the House of Commons.

After the construction of County Hall in 1922, Westminster Bridge provided a link over the Thames between the home of the national government and that of London's government, until the abolition of the GLC in 1986. At either end of the bridge are sited two distinctive sculptures. On the County Hall side stands a 13-ton lion made in 1837 of Coade stone. Coade stone is an artificial stone made of terracotta mixed with ground quartz and glass to make it extremely long-lasting and waterproof. The firm which made it was located on the site of County Hall until it closed down in 1840. The so-called Coade Lion was originally painted red and stood over the entrance arch of the Red Lion Brewery, which was located by Hungerford Bridge near today's Festival Hall. The brewery was demolished in 1949, but the lion reappeared at the Festival of Britain in 1951. When the festival closed, the lion was saved by the personal intervention of George VI, and it was moved to Waterloo Station. In 1966, when Waterloo Station was redeveloped, it was decided to re-erect the lion on the empty pedestal by County Hall. Its present colour is white, although the real colour of Coade stone is tawny yellow. Unseen to the public is a small room underneath where security guards can make themselves a cup of tea.

The bronze sculptural group on the opposite side of the bridge is of Queen Boudicca with her two bare-breasted daughters, driving a scythed chariot in full battle cry. The sculptor was Sir Hamo Thorneycroft, who specialised in heroic sculptures of famous Britons. His statue of Oliver Cromwell stands near by outside Westminster Hall. Strictly speaking, Boudicca should be driving her chariot through the City rather than through Westminster, since it was there that she slaughtered the Romans before her final defeat and death at their hands. This happened long before Westminster was founded. An amusing letter appeared in *The Times* of 25 August 1964 concerning a court case in which a young woman was charged with indecency for wearing a topless dress on Westminster Bridge. The writer pointed out that: 'The Magistrate failed to take account of two other women in topless dresses seen hanging about on the north bank of the bridge for some little time now in a chariot driven by their mother.'

Old Westminster Bridge, completed in 1750 in the classical style, with its 13 Portland stone arches, was a fitting

Westminster Bridge's Coade stone lion

embellishment to London's Georgian Age. The present bridge has stood the test of time much better, although the scour protection for the central piers had to be renewed in 2005. Unlike the old bridge, it could never quite aspire to be an emblem of London, since it is dominated by the Houses of Parliament, whereas its predecessor had only to compete with the ramshackle collection of buildings that comprised the Palace of Westminster in the eighteenth century. Over the past two and a half centuries of their joint existence, the two bridges have witnessed many colourful events. Until 1882, the Lord Mayor's procession came by river, and the state barge landed at Westminster Bridge for the swearing-in ceremony in Westminster Hall. The pageant inspired many artists, including John Constable, to paint the scene. The annual London-to-Brighton veteran car race crosses the river at Westminster Bridge from its starting point in Hyde Park. The race was initiated in 1896 to celebrate the abolition of the law compelling motorists to be preceded by a man with a red flag, and the raising of the speed limit from 4 mph to 14 mph. On the night of 31

December 1999, the crowds on Westminster Bridge watched in anticipation as the hands of Big Ben moved inexorably towards twelve midnight GMT, and the clock's massive bell chimed in the new millennium.

Today, crowds mass on the bridge to see a very different skyline from that which inspired Wordsworth's sonnet. Although the open fields are long gone, the view is still inspiring, with the massive circle of the London Eye downstream and the Houses of Parliament upstream. About the only structure remaining from Wordsworth's time is the Portland stone of the bridge's abutments, but sadly this is faced with brick, so no one ever sees it.

CHAPTER 9

Charing Cross

A combined railway and pedestrian bridge crosses the river today from Charing Cross on the north bank to the Royal Festival Hall on the south. The railway bridge was constructed in 1864, replacing an earlier suspension bridge of 1845 which catered only for pedestrians. There was a rather forbidding footpath right next to the railway on the downstream side, but in 2002 this was replaced by two new cable-stayed footpaths which are attached at a comfortable distance from the railway bridge. It has always been known as Hungerford Bridge after the family who owned a mansion and founded a market here on the north bank. The new footbridges were given the collective name Golden Jubilee Bridge in honour of the Queen's Jubilee in 2002.

Charing Cross is considered to be the centre of London, from where all mileages to remote towns in Britain are measured. It is therefore surprising that no bridge was built here until 1845. One of the reasons for this is its location on the greatest curvature of the sharp bend at this point on the river. Indeed, the name Charing Cross comes from the Saxon word 'char', which means 'bend in a river'. There is no truth in the more romantic idea that the name comes from '*Chère Reine*', referring to Eleanor of Castile, the beloved wife of Edward I, for whom he erected crosses at the 12 places where her funeral cortège stayed on its way from Harby in Nottinghamshire to Westminster Abbey. The last cross was sited where today there is a plaque indicating the exact spot from where all mileages are measured, close to the equestrian statue of Charles I in Trafalgar Square.

The place has been at the centre of events crucial to the nation's history, one of which involved the Earl of Pembroke, an ancestor of the earl who had supported Labelye throughout the Westminster Bridge project. The Earl commanded the royal troops in opposition to the rebel forces of Sir Thomas Wyatt, who in 1554 tried to overthrow the Roman Catholic Mary I. After a confused battle in what was then the small village of Charing, today the area around Trafalgar Square, the rebels surrendered and Sir Thomas was executed after implicating Mary's sister, Elizabeth, in the rebellion. Elizabeth herself survived imprisonment in the Tower and became Queen of England after Mary's death.

A century later, the reigning monarch, Charles I, was not so fortunate. He was defeated in the Civil War and executed on a balcony outside the Banqueting House in Whitehall in 1649. His bronze equestrian statue, which was designed by Hubert Le Sueur and now stands at the Trafalgar Square end of Whitehall, was cast in 1633. After the King's execution, the statue was sold to a brazier named Rivers to be broken up. Rivers evidently carried on a brisk trade in bronze cutlery which he claimed was part of the melted-down statue. However, he had in fact buried the statue and it was re-erected at its current location after the restoration of the monarchy in the person of Charles II in 1660. Several of the people responsible for the execution of Charles I, the new king's father, were themselves executed on this spot, as recorded by Samuel Pepys in his diary entry for 13 October 1660:

> I went out to Charing Cross to see Major-General Harrison hanged, drawn and quartered . . . Thus it was my chance to see the king beheaded at Whitehall and to see the first blood shed in revenge for the king at Charing Cross.

It was Charles II who in 1679 gave permission to Sir Edward Hungerford to hold a market at Charing Cross. From medieval times, bishops and aristocrats built mansions on the riverside with their fronts facing the Strand and with gardens running down to the riverside where watergates provided access for their private barges. One of these mansions, located on the site of today's Charing Cross Station, belonged to the Hungerford family. According to *The Survey of London*, 'the family record for dishonesty, vice and violence seems to have been exceptional even in the unsqueamish age in which they flourished'.[26] The record of violence was begun in the fourteenth century by Agnes Hungerford, who had been married to John Cotell, the steward of one of Sir Edward Hungerford's ancestors, also called Edward. Agnes decided to better herself when she realised that Sir Edward could not keep his eyes off her. She persuaded a couple of yeomen to strangle her husband and hide the crime by throwing his body on to the hearth of the kitchen at Farleigh Castle, the Hungerford's country estate. Sir Edward then married her but died soon afterwards, having bequeathed her the majority of his inheritance. Unfortunately for her, Sir Edward's son by his first wife was so furious at being excluded from his father's will that he informed on his stepmother and this resulted in her execution at Tyburn in 1523.

The later Sir Edward, who founded the market in his family name, was the last in line of the Hungerfords. He died in 1711 at a great age, having squandered the family fortune. It is said that the original market building was designed by Sir Christopher Wren. By the nineteenth century, the market had lost out to Covent Garden and fallen into disuse. According to the *Survey*, Hungerford Market was 'little better than a monster dust-heap and a cemetery for the dead dogs and cats of the neighbourhood'. Near by was sited the blacking factory where Charles Dickens worked as a boy, putting labels on tins of boot polish. His experiences there are described in his semi-autobiographical novel *David Copperfield*, in which he wrote: 'No words can express the agony of my soul as I sank into this companionship and felt my hopes of growing up to be a learned and distinguished man crushed in my breast.'

In 1830, a new Hungerford Market Company was set up to take advantage of the waterfront location, which provided cheap river transport for fish and other commodities. Charles Fowler, who had just constructed the new market buildings at Covent Garden, was commissioned to design a large hall to house the market stalls. Fowler's new hall had an impressive porticoed frontage facing the river, and the roof was raised on tiers of open arches which provided light and air to the inside of the building. The market was opened in 1833 with great ceremony, including the ascent of a hot-air balloon.

With every expectation that the exciting Italianate architecture and riverside location would lead to the success of the market, a private Act of Parliament was obtained authorising the construction of a suspension footbridge from Hungerford Market to Lambeth, to provide access for the inhabitants of the south bank, where there was no fish market. The bridge was designed by Isambard Kingdom Brunel. The river was 1,350 feet wide at this

Brunel's Hungerford Suspension Bridge of 1845

point, and the bridge provided a 14-foot-wide footway suspended from two campanile towers supported by solid masonry river-piers clad in ornamental red brickwork. The central span of 670 feet was longer than any other bridge in Britain, including Telford's Menai Bridge, and was only exceeded by the suspension bridge at Fribourg in Switzerland. The bridge was officially opened as the Charing Cross Bridge in 1845, but has always been known as Hungerford Bridge because of its association with the market. The total cost was £100,000, and the company aimed to make a profit from its investment by charging a halfpenny toll. Unfortunately, as stated in the *Illustrated London News* of 3 December 1842, the impressive bridge provided communication to the 'worst parts of Lambeth', and the profits never materialised. Moreover, Fowler's market hall was burnt down by a fire in 1854, resulting in the demise of the market.

Isambard Kingdom Brunel (1806–59)

Isambard Kingdom Brunel was born in Portsmouth, the eldest son of Marc Isambard Brunel and Sophie Kingdom, from whom he inherited his distinctive first names. In 1823, he was apprenticed to his father, who had started constructing the first-ever tunnel under a major river, between Wapping and Rotherhithe. He was appointed resident engineer for the project in 1826, but the following year almost proved fatal when one of several inundations occurred. Brunel, who was working in the partly excavated tunnel at the time, was swept away by the rushing water and only just managed to escape drowning. During his convalescence from his injuries in Clifton, he entered a competition for the design of a bridge over the Avon Gorge. Despite opposition from the doyen of British bridge builders, Thomas Telford, Brunel was awarded the contract for the Clifton Suspension Bridge.

In 1833, the Great Western Railway appointed Brunel as chief engineer at the young age of 26, and he devoted the next 15 years to the massive project of constructing the railway from

Paddington to Bristol Temple Meads Station. During his work on the railway, he persuaded the GWR to extend its route to New York by building the SS *Great Western*, which was the largest steamship of its day. He followed this by constructing the SS *Great Britain*, which was the largest iron ship and the first large screw-propelled ship in the world, and then the massive SS *Great Eastern*, which at 32,000 tons was so large that it required several attempts to launch it. This last project exhausted Brunel, who had a seizure on deck shortly before the maiden voyage and died a few days later.

Both Brunel's achievements and disasters were on an epic scale. He is widely acknowledged to have been the greatest of the many outstanding Victorian engineers and was voted second-greatest Briton after Sir Winston Churchill in a BBC poll in 2002. He is buried in Kensal Green Cemetery, and his statue by Marochetti stands on the Victoria Embankment next to Somerset House.

Isambard Kingdom Brunel statue, located near to Somerset House

In 1854, a proposal was made to widen Hungerford Bridge so that it could be used for the transport of vehicles, but this was rejected in favour of the South Eastern Railway (SER) plan to run its trains across the river to a station at Charing Cross. The SER obtained authorisation by an Act of Parliament in 1859 to purchase Brunel's suspension bridge and replace it with a railway bridge linking a new station at Charing Cross to the existing station at Waterloo.

There was fierce competition between the rival railway companies for access from the south to termini on the north bank. The London, Brighton and South Coast Railway had already crossed the river to Victoria Station, and the London, Chatham and Dover Railway was about to do the same at Blackfriars. Not everyone approved of these schemes. As pointed out in the *Illustrated London News* of 21 February 1863:

> The first Railway Acts stipulated that no locomotives should come into the streets of London proper. Now this is to be changed. Locomotives are to be allowed to career about and under the thoroughfares of London pretty much at the discretion of engineers and directors.

The strongest opposition came from St Thomas's Hospital, which demanded and received £250,000 compensation because the railway line to Charing Cross passed too close to its site near London Bridge. The hospital was forced to relocate to its present site opposite the Houses of Parliament.

Sir John Hawkshaw designed the new Charing Cross Bridge and the station. The bridge was formed of wrought-iron lattice girders, providing four railway lines across the Thames, which was then 1,350 feet wide at this point. To support the bridge girders, Hawkshaw used the bases of Brunel's original brick-clad river-piers, as well as pairs of massive cast-iron cylinders sunk 30 feet into the clay bed of the river and then filled with concrete. The result, according to the Institution of Civil Engineers *Proceedings* of 1862–3, was the strongest and cheapest bridge of its kind in the world: 'Hawkshaw succeeded in designing a

Hungerford Railway Bridge of 1864 with one of Brunel's old river-piers

bridge that is well adapted to the situation where it is placed and for the purpose intended and its comparatively small cost proves that no outlay has been incurred beyond what was needed.' Mr Hemans, one of the members present at the meeting, stated that he 'regarded this bridge as one of the most perfect pieces of ironwork of the kind ever produced'.

John Hawkshaw (1811–91)

Hawkshaw was born in Leeds, the son of a publisher. He became interested in railways at an early age and worked on many railway and canal projects in the north of England. He achieved notoriety when in 1838 he attacked the broad gauge used by Brunel for the formidable GWR, and in this he was proved right. In 1850, he came to London to set up his own practice as a consulting engineer. Although he worked as chief engineer for the SER from 1861 to 1881, he did not confine himself to railway engineering. His knighthood in 1873 resulted from his design of Holyhead Harbour, and his many other schemes included sewage systems, land drainage and flood prevention. In London, apart from Hungerford Bridge, he designed railway bridges over the Thames at Cannon Street and Staines, and the southern dock basin for the West India Dock Company, as well as converting Marc Brunel's pedestrian Thames Tunnel for use by the East London Railway. Hawkshaw's reputation quickly spread abroad. He advised the Viceroy of Egypt on the Suez Canal proposal, designed the 16-mile Amsterdam Ship Canal and worked on projects as far apart as Mauritius, Jamaica, India, Russia and Hungary. Towards the end of his career he even worked on a proposal for the abortive Channel Tunnel Company.

The praise initially lavished on Charing Cross Bridge, as it was first called, has not been replicated by later critics or the public. M.P. Burke produced a paper, published in the Institution of Civil Engineers *Proceedings* of February 1998, in which he named Hawkshaw's Hungerford Bridge as one of the five most aesthetically notorious bridges in the world. The other four were the Lansdowne Bridge over the River Indus, the Queensborough and Williamsburg bridges over New York City's East River, and Tower Bridge. The author had arrived at these five nominations following detailed research into all the literature on bridge aesthetics published during the previous 100 years. One of the conclusions reached by the author was that Hungerford Bridge disproves the often-heard fallacy that if a structure is really scientifically designed it must be beautiful.

The only person who seems to have been able to find beauty in the rectilinear lines of the bridge's metal framework was Claude Monet. Monet stayed at the Savoy Hotel in London on several occasions from 1899 to 1901 and had a view of Hungerford Bridge from the window of his room on the sixth floor. He painted no fewer than 35 canvases showing the bridge under different atmospheric and light conditions. The effect of air pollution, combined with the billowing steam of the crossing trains, allows the straight lines of the bridge to merge and almost evaporate into the surrounding haze in these evocative Impressionist paintings. It is doubtful if Monet would have painted the scene with such enthusiasm in the clearer light of today.

As for Brunel's original suspension bridge at Charing Cross, the red-brick piers were not the only remains to be preserved. Brunel had designed a suspension bridge over the Avon Gorge at Clifton near Bristol but the money ran out in 1843, soon after work had begun. In 1860, the SER agreed to sell the old Hungerford Bridge suspension chains for £5,000 to a consortium which aimed to complete Brunel's design after his death. After some modification to allow the chains to carry a vehicular roadway instead of the pedestrian footway for which they were originally designed, the consortium, of which Hawkshaw was a member, went on to create what is today the most famous structure remaining of all the works of the great engineer. The Clifton Suspension Bridge was finally opened in 1864, the same year as Hungerford Railway Bridge.

Although the cost of Hungerford Bridge itself was only £180,000, the SER had to spend a total of four million pounds to extend the railway from Waterloo to Charing Cross. To help recoup some of this huge investment, footways were constructed on the cross-girders which extended to both sides of the railway, and tolls were charged for pedestrian crossings until the Metropolitan Board of Works bought out the SER's right to levy tolls for £98,000 in 1878. The upstream footway was removed in 1886 when Hungerford Bridge was widened to carry an extra four tracks. An additional source of revenue was generated from the steamboat landing stage which was constructed at the base of the southern red-brick pier and was accessible via stairs descending from the footpath.

Sir Joseph Bazalgette's scheme for embanking the Thames and carrying London's sewage out to the east to be emptied into the Thames at Beckton and Crossness was begun in 1862, just before Hungerford Bridge was completed. At this point in the river, the Victoria Embankment runs underneath the bridge where it fans out to take the railway tracks into the station platforms. The resulting reduction in the width of the river is highlighted if we note the site of the sixteenth-century structure known as York Watergate which stands in Victoria Embankment Gardens, where the river used to flow before the embankment was built. Bazalgette's bust, designed by the sculptor George Simonds and inscribed with the words 'Engineer of the London Main Drainage System and of this Embankment', is located a few yards upstream of the bridge.

Joseph Bazalgette (1819–91)

Like Isambard Kingdom Brunel, his friend and fellow engineer, Joseph Bazalgette came from a family of French immigrants. He took an early interest in engineering and was apprenticed to the firm of the respected engineer Sir John MacNeill, where he gained considerable experience in land drainage and reclamation while working on the railways. In 1842, he was able to set up his own consulting practice.

The Metropolitan Commission of Sewers was set up in 1847 with the aim of improving London's sewage system. Its first edict was to order that all the cesspits should be closed, with the result that the excreta of London's total population of three million was disgorged directly into the Thames. Bazalgette was appointed assistant surveyor to the Commission in 1849, by which time over 14,000 people had died in a cholera epidemic induced by drinking polluted water. In 1856, the Commission was abolished. It was immediately replaced by the MBW, and Bazalgette obtained the job of chief engineer following an enthusiastic recommendation by Brunel. The hot summer of 1858 resulted in what became known as the 'Great Stink', when the Members of Parliament were overwhelmed by the stench from the polluted river and were induced to pass legislation enabling the construction of a radical new drainage system for

London. Bazalgette designed the system of embankments and intercepting sewers which solved the problem of cholera and is the basis of London's present sewage network.

He received a knighthood in 1875, and was elected president of the Institution of Civil Engineers in 1888. Apart from his great work on London's sewers, Bazalgette was responsible for many other civic works, including the construction of Northumberland Avenue, leading from Trafalgar Square to the Thames by Hungerford Bridge. He also designed three of London's river crossings, at Hammersmith, Putney and Battersea, all of which survive.

The railway line crossing Hungerford Bridge was and still is popular with Londoners because it takes passengers directly to the very centre of town. To complement the station, E.M. Barry, the son of the architect of the Houses of Parliament, Sir Charles Barry, constructed the Charing Cross Hotel, with its French-Renaissance-style Strand frontage, together with the replica of the Eleanor Cross which stands in the forecourt. This impressive location proved so popular that 'under the clock at Charing Cross' became a favourite meeting place. According to Volume 75 of *Railway Magazine*, 'Half feminine London used to wait there at night for its young man, and the other half said that was who it was waiting for.'

The atmosphere on the riverbank by Hungerford Bridge, where the trains approached the station, was less salubrious. Here Rudyard Kipling lived for a time as an unknown writer on his return from India in 1889. He chose this location because of the low-priced accommodation and cheap food obtainable in the nearby eatery at the bottom of Villiers Street, which is known today as Gordon's Wine Bar. The fumes from the railway inspired him to write his first novel, *The Light That Failed*.

In 1926, the Royal Commission on Cross-river Traffic made wide-ranging recommendations on London's river bridges. Regarding Hungerford Bridge, the Commission proposed a new combined road/rail crossing, as well as the widening of Waterloo Bridge in order to cater for the increasing road traffic in this central area of London. Neither of these proposals was implemented. Then, in 1930, London County Council submitted a Bill requesting authorisation to transfer Charing Cross Station to the south bank of the river on the site of the Lion Brewery and to construct a new road bridge in place of Hungerford Railway Bridge. Parliament rejected the Bill on the grounds of the high estimated cost of 13 million pounds and the unsuitability of the site for the station.

Nevertheless, public pressure for a road bridge continued to build, and the City of Westminster instigated a review of the Charing Cross Bridge schemes with a view to making its own proposals. The report, published on 31 May 1935, found that traffic was indeed growing, partly because of the shift from horse-drawn vehicles to motorised transport. In 1924, 17 per cent of traffic in central London was horse-drawn, but by 1933 this had reduced to 1.5 per cent. It was concluded that a new road bridge at Charing Cross was highly desirable but that the cost would be considerably more than the 13 million pounds estimated by the LCC, and it would only be possible to go ahead if the Government contributed most of the money. Unsurprisingly, nothing was done and Hungerford Bridge survived until the Second World War, when enemy bombs nearly destroyed it. In 1941, bombs did severe damage to the Charing Cross Hotel and set fire to four trains in the station. A German landmine came to rest near the signal box on the bridge, and its parachute gear became entangled in the bridge girders.

Having survived the onslaught of town planners and enemy bombs, Hungerford Bridge was again the centre of attention when the South Bank was transformed for the 1951 Festival of Britain. The aim was to mark the centenary of the 1851 Great Exhibition with a celebration of British achievements in the arts, architecture, science and technology. The area on the south of the river by Hungerford Bridge had been derelict following the demolition of the Lion Brewery in 1949. About the only remaining structure was the tower of a former lead-shot factory.

The 1951 Festival of Britain site under construction,
with the temporary footbridge between
Westminster and Hungerford Bridge

The exciting modern architecture set up here for the Festival, including the Skylon obelisk, the Dome of Discovery and the Royal Festival Hall, thrilled the nation, accustomed to the dreary building standards employed for the massive housing developments required after the end of the Second World War, when rationing had still not been abolished.

The main concept of the Festival was to portray the British contribution to civilisation as springing from the combination of two forces – the initiative of the people and the resources of their native land. Since Hungerford Bridge crossed the river at the mid-point of the South Bank site, it became the dividing agent between the two facets of the exhibition. The narrow footpath on the downstream side of Hungerford Bridge was totally inadequate to carry the massed crowds of visitors coming from central London, and therefore a Bailey-type steel footbridge had to be constructed on the upstream side of the bridge for the duration of the Festival. This footbridge was demolished in January 1953 after the removal of all the South Bank buildings apart from the Royal Festival Hall.

Although popular with passengers for its convenient central location, Hungerford Bridge and Charing Cross Station remained a target for town planners and architects. In 1986, Richard Rogers put forward an ambitious scheme reminiscent of the 1930 LCC plan. He proposed removing the station entirely and whisking passengers across the Thames from Waterloo Station on a futuristic cable railway. In place of the station he envisaged a riverside park under which the Embankment roadway would run in a tunnel. Shortly after this, Terry Farrell was commissioned to design an air-rights development above the station, and all thoughts of radical solutions such as those of the LCC and Richard Rogers have been abandoned. Farrell's Embankment Place rises dramatically over the station with its two vast glazed barrel vaults enclosed by four towers of Sardinian granite. Hawkshaw's original station fronted the river with a great arched glass-and-iron train shed, but this collapsed in 1905 as a result of a fire in which five people died. Farrell's postmodern design certainly presents an equally striking façade to the river, as well as providing profitable office space.

The latest stage in the Hungerford Bridge story has been the construction of the Golden Jubilee Bridge, consisting of two footbridges on either side of the railway. Lifschutz Davidson won the 1996 design competition sponsored by the Cross River Partnership, which was a consortium of local authorities and Railtrack. The contract was awarded to Costain/Norwest Holst. Each footbridge is 4.7 metres wide and is supported by painted white steel rods fanning out from steel pylons, which slant outwards as they rise from concrete piers which are embedded far into the river-bed. The two footbridges replaced the single, narrow seven-foot-wide crossing which was previously used by three and a half million pedestrians annually. According to an article in *The Times* of 13 April 2000 about the inadequacies of the old footbridge, pedestrians 'ran a gauntlet of beggars and drunks as they crossed from Charing Cross to South Bank'. The south bank stairs were said to be the most lucrative begging pitches in London, where it was possible to earn fifteen pounds in a half-hour slot from concert-goers on their way to and from the Royal Festival Hall.

The footbridge was the scene of a brutal murder of a student by a gang of six teenagers in the early hours of the morning on 18 June 1999. Timothy Baxter was crossing the bridge with his friend Gabriel Cornish when they were attacked and robbed by three youths. Mr Cornish saw two young men and a girl approaching from the other direction and called out for help. It turned out that the new arrivals were part of the same gang, and they joined in beating up the two friends. One member of the gang then called out that 'it would be fun' to throw them into the Thames after they were beaten unconscious. The gang hauled them up over the railings and let their unconscious bodies fall into the river. Amazingly, Mr Cornish was rescued, but Mr Baxter drowned and his corpse was found in the river the following morning. The gang were caught on CCTV and were seen joking together after they had thrown the bodies into the river. They tried to blame each other for Timothy Baxter's death, but a jury found they were all responsible for his killing and the attempted murder of Cornish. The judge sentenced them all to life imprisonment.

Security on the Golden Jubilee Bridge is provided by 18 CCTV cameras, and the balustrade curves away from the footpaths so that it is much harder to jump or be thrown into the river than was the case with the low, upright railings on the old footbridge.

The original plan was to open the footbridges in the year 2000 to coincide with the millennium, but the project was delayed because of objections raised on safety grounds by London Transport (LT). LT had initially backed a rival scheme for a foot tunnel under the Thames along the same route. Since the river-piers of the new bridges were to be driven 42 metres into the river-bed, LT was concerned that the necessary piling could set off Second World War bombs which might still be embedded at this point. As described above, the bridge had been subjected to severe bombing in 1941, and one unexploded bomb was found in the river in the 1950s. The fear was that an explosion could cause the Thames to flood into the whole Underground system via the Northern Line tunnel which runs under the river next to Hungerford Bridge, resulting in a catastrophe. LT had identified a real risk, but a solution was finally found. Since the greatest danger was at the Charing Cross side of the bridge, the final pylon on the northern end was made to rest on the north bank instead of on a river-pier. This required a longer span at this point and significant redesign of the pylon.

A further problem arose when the Millennium Bridge between Tate Modern and St Paul's Cathedral was opened on 12 June 2000 and was immediately closed because of the 'wobble' induced by the large numbers of people crossing over it. Consequently, questions were asked about the rigidity of the

Golden Jubilee Bridge

Golden Jubilee Bridge design. Engineering calculations showed that there would be a problem only if a group of vandals did simultaneous rhythmic jumping in order to deliberately cause a wobble, so it was decided not to incur the considerable cost of installing dampers. Finance had run out by this time, and the footbridges might have remained unfinished if the Greater London Authority had not agreed to provide a grant of £16,700,000 towards the total cost of £39,500,000. The upstream footbridge was finally opened in May 2002, in time for the 50th anniversary of the Queen's accession, and hence the structure was named the Golden Jubilee Bridge. The downstream bridge was opened in the following September.

The delicate tracery of the slim white rods of the footbridges provides a dramatic contrast to the massive horizontal girders of Hawkshaw's railway bridge, which they enclose but do not hide. The effect is especially appealing at night, when lights concealed in the tops of the pylons shine down the white rods as they fan out to illuminate the deck of the footbridge below. The combination of the twenty-first-century Golden Jubilee Bridge with the nineteenth-century Hungerford Railway Bridge may not be a perfect aesthetic or practical solution for a river crossing at this prestigious location, but it is certainly unique. Few people today would number it among the five ugliest bridges in the world.

CHAPTER 10

Waterloo

The present Waterloo Bridge, constructed in reinforced concrete, was completed in 1944 towards the end of the Second World War. It replaced John Rennie's internationally admired stone bridge, which was opened in 1817.

The original name of the bridge was Strand Bridge, referring to the Strand, which runs parallel to the north bank of the Thames in this part of London. Today, the Strand is separated from the river by substantial buildings and the Victoria Embankment, but in earlier times, as implied by the name, it was a track running along the shore of the much wider river. It has been known for some time that a Saxon settlement must have existed by the Strand. In AD 732, the Venerable Bede wrote of 'an emporium of many people coming from land and sea' to conduct trade in the area. In 1884, archaeologists discovered the so-called Waterloo Bridge Hoard, consisting of coins buried in AD 875, at what is today the Charing Cross end of Hungerford Bridge. In the 1980s, detailed archaeological excavations in the area of Covent Garden established for a fact that the Saxon settlement of Ludenvic was sited in this area on land sloping down to the river, until the ninth century, when Alfred the Great rebuilt Roman London as a new Saxon burgh. Soon afterwards, the settlement of Ludenvic became derelict, but it is still commemorated in the name Aldwych, which means 'old town'.

From the twelfth century, aristocrats and bishops were attracted to build their London residences in the area because it was conveniently located between the Palace of Westminster and the City of London. The Strand also offered a river frontage from which they could embark in their private barges to avoid the unpleasantness of travelling along the muddy, crowded and dangerous roads. Two of the most impressive riverside mansions were the Savoy Palace and Somerset House, which were located on either side of the future Waterloo Bridge.

The Savoy Palace no longer exists and is remembered only in the names of the present theatre and hotel. The palace's origins date back to the thirteenth century, when Henry III granted land to Peter de Savoie, the uncle of Queen Eleanor. Peter constructed a house here in his family name in 1263. The original house was considerably expanded into a palace by later owners, the most illustrious of whom was John of Gaunt, Duke of Lancaster and virtual ruler of England during the boyhood reign of Richard II. Gardens ran down to the Thames where today Lancaster Place forms the approach to Waterloo Bridge. William Chaucer, who as a young man was one of John of Gaunt's retainers and was married to his sister-in-law, wrote a poem entitled *The Boke of the Duchesse* in honour of the Duke's first wife, in which the following lines are thought to refer to the Savoy Palace rose gardens:

A garden saw I full of blossomed bowls
Upon a river in a garden mead.[27]

Soon after Chaucer wrote these lines, the Savoy Palace was attacked and burned down in the 1381 Peasants' Revolt. Evidently the attackers threw all the furnishings and valuables into the Thames, convinced that no one should own anything belonging to such an evil man as John of Gaunt. There is no record of what happened to these treasures – certainly they had long vanished when Waterloo Bridge was constructed near the site of the old palace in the nineteenth century. The Savoy

Palace was soon rebuilt after the suppression of the Peasants' Revolt, but by the sixteenth century it had degenerated, and on his death in 1509, Henry VII left instructions in his will for it to be turned into a hospital providing nightly lodgings for poor men. Ironically, only the very rich can now afford to reside in its successor building, the Savoy Hotel. By the beginning of the nineteenth century, the old Savoy buildings had become ruinous. The area was therefore cleared for the construction of the approach roads for Waterloo Bridge. The only remaining structure is the sixteenth-century Chapel of the Savoy, located in Savoy Hill.

Somerset House was originally built in 1551 as a Renaissance palace for the Duke of Somerset, Lord Protector of England during the reign of the boy king Edward VI. Somerset was executed soon after completion of his palace, which was then handed down to a succession of princesses and queens until it was demolished in 1775, to be replaced by the present building. The architect of the new Somerset House was William Chambers, who designed the imposing classical complex for use as offices to cater for the expanding bureaucracy of the government of Britain and its colonies. The river frontage, with its heavily rusticated Doric columns and impressive watergate, is today separated from the Thames by the Victoria Embankment but at that time could be approached by boat. It was to inspire the architecture of the first bridge at this location.

The opposite side of the river has a less distinguished history. The land was marshy and largely uninhabited until wharves and industrial enterprises were set up along the riverbanks in the eighteenth century. The area was known as St George's Fields after the church of St George the Martyr in Borough High Street. According to Thomas Pennant, it was renowned for its vineyards, which produced excellent white wine and vinegar.[28] When Westminster and Blackfriars bridges were built in 1750 and 1769 respectively, the roads leading south from them converged at St George's Cross, known today as St George's Circus. The southern approach road to Waterloo Bridge was about to join them at this point.

For centuries, people crossed the river here, as elsewhere in London, by wherry from the numerous stairs where watermen waited to attract customers who did not own their own barge. By 1800, the population of London exceeded one million. London, Blackfriars and Westminster bridges were the only permanent crossings and proved totally inadequate to handle the increasing traffic. In 1806, a group of speculators formed the Strand Bridge Company with the aim of constructing a bridge across the Thames midway between Westminster and Blackfriars. As usual, vested interests opposed the idea, but Parliament had heard similar objections before and was unlikely to reject the idea completely. Several members of the House of Lords banded together to propose that a wooden bridge should be built initially, with a view to establishing the real demand. The income from the tolls could then be used to construct a stone bridge later if necessary. The group was mocked by the press and given the nickname of 'the wooden peers'. Its proposal was not implemented.

In 1809, the Strand Bridge Act was passed, authorising the construction of a stone bridge to be financed by the income from tolls. The sum of £500,000 was raised without any difficulty and the shares immediately reached a premium of one guinea. However, the initial enthusiasm soon subsided when additional finance was required. The final cost of the new bridge, including its approach roads, amounted to £937,000.

George Dodd, son of the unfortunate Ralph Dodd, the proposer of a number of unsuccessful bridge projects, submitted the design. A pioneer of steam navigation, he had built a steamship called *The Thames*, which took passengers to and from Margate. He does not seem to have had any experience of bridge design, although he may have relied on his father's advice. His design was closely based on the Pont de Neuilly, which had been constructed 40 years before over the River Seine in Paris by the great French bridge builder Perronet. The Strand Bridge Company asked the engineers John Rennie and William Jessop for their opinions of the design. They disapproved of the idea of constructing a mere copy of a French bridge over this historic central point on London's river and made a number of specific criticisms. They especially objected to Perronet's *cornes de vache*, which were designed to make the bridge spans look flatter than they really were by slanting the fronts of the arches inwards to the base of the piers. Dodd's design was therefore rejected.

Rennie submitted two designs of his own, one for seven and one for nine spans. The company selected the nine-span design on grounds of cost. Rennie's bridge consisted of nine 120-foot semi-elliptical river arches of Cornish granite. The total length, including the abutments, was 1,380 feet. The plain surface of the arches was broken by incorporating three-quarter Doric columns over the piers to support projecting recesses on the parapets. Rennie's stated aim was to do justice to Somerset House, the Palladian arched frontage of which was lapped by the river on the downstream side of the bridge. By linking the bridge to York Road on the south bank over arched approach roads, Rennie managed to provide a virtually flat crossing to the Strand and at least in this respect he did emulate Perronet's Pont de Neuilly.

The bridge architecture was greatly admired. The Italian sculptor Canova called it 'the noblest bridge in the world' and said that 'it is worth going to England solely to see Rennie's bridge'. This was a remarkable accolade at a time when it was more normal for English aristocrats to travel to Italy on the Grand Tour for their cultural education. This architectural excellence led Sir Reginald Blomfield to question whether Rennie, a mere engineer, could have designed the bridge on his own, especially as his London Bridge was inferior from an architectural point of view.[29] Ralph Dodd also challenged Rennie's authorship of the design. Writing in *The Gentleman's Magazine* of June 1817, he claimed that his family in the person of George Dodd was the true designer and offered to show the original plans to prove it. There is, however, no evidence to support either of these critics. Moreover, Rennie's 1803 design for Kelso Bridge in the Scottish Borders forms the basis for his bridge here on the Thames.

John Rennie (1761–1821)

John Rennie was the fourth son of a Scottish farmer who lived near the village of East Linton, a few miles east of Edinburgh. He took an early interest in machinery and when only twelve years old he left school to work for Andrew Meikle, the distinguished millwright. Although he later returned to full-time education, including attending lectures by Professor Robson at Edinburgh University, he was determined to pursue a career as an engineer. In 1783, Rennie journeyed south to London, where through the recommendation of Professor Robson he obtained a job with James Watt as a millwright. His first major commission was the construction of the steam-powered machinery for Albion Mills at the southern end of Blackfriars Bridge. The mills were soon destroyed by arson because their superior automation threatened competitors and jobs. However, Rennie's achievement established his reputation and enabled him to set up his own practice.

He quickly expanded his interests and worked on increasingly large-scale projects in canal and bridge building, in improving the drainage of the Fens in Norfolk and in the construction of docks. His major projects included the Kennet and Avon Canal, the London Docks in Wapping and the great breakwater at Plymouth, which created the Royal Navy's premier harbour. His three bridges over the Thames expanded the state of the art in terms of the width and flatness of their arches. They all lasted for over 100 years, but eventually had to be replaced. Rennie died in 1821 before work started on his final project at London Bridge. He was ranked with Telford, Stephenson and Brunel as among the greatest civil engineers of his age and was buried in the crypt of St Paul's Cathedral near Sir Christopher Wren.

Construction started with the laying of the foundation stone of Cornish granite on 11 November 1811. A set of contemporary coins enclosed in a glass case was placed underneath the foundation stone. Crowds flocked to watch progress, among them Tsar Alexander I of Russia, who visited the site several times on his state visit to London in 1814. Rennie employed a number of revolutionary techniques during the project, including the first use of steam pumps to remove water from the cofferdams which were used to build the river-piers. The bridge's granite facings were transported from Aberdeen and Cornwall, but the remainder of the stone was hewn in nearby fields on the south bank. Samuel Smiles tells the story of how the stone was transported by truck on temporary rails drawn by horses.[30] Most of these trucks were drawn by Old Jack, a popular local carthorse. His driver would stop for a drink with his friends at an inn by the track, and often the drinking time was prolonged. On one occasion, Old Jack became impatient, poked his head in at the open door of the inn, took his master's coat-collar between his teeth and dragged him out to resume working.

Smiles also states that Rennie surfaced the bridge roadway using the same method patented by John Loudon McAdam six years later. Instead of laying the gravel and flint on the roadway in its natural state, which allowed vehicle wheels to churn up the surface and leave large ruts, Rennie first levelled off the gravel and then covered it with broken flint, which he pressed down into the gravel to form a firm 'macadamised' surface. Had Rennie taken the credit due to him, we would have had to use the term 'tarren' instead of 'tarmac'.

The year after Tsar Alexander's visit saw the momentous victory of the Allies against Napoleon at Waterloo. Patriotic fervour was such that the Strand Bridge Company renamed itself the Waterloo Bridge Company following a new Act of Parliament of 1816. The Act explained the change of name in unusually florid language for a legal document:

> Whereas the said bridge when completed will be a work of great stability and magnificence, and such works are adapted to transmit to posterity the remembrance of

great and glorious achievements: and whereas the said
Company are desirous that a designation should be given
to the said bridge which should be a lasting record of the
brilliant and decisive victory achieved by His Majesty's
forces, in conjunction with those of his allies, on the
eighteenth day of June one thousand, eight hundred and
fifteen; Be it therefore further enacted, That from and
after the passage of this Act the said bridge shall be called
and denominated the Waterloo Bridge.

The old Waterloo Bridge of 1817

Waterloo Bridge was opened by the Prince Regent,
accompanied by the Duke of Wellington, on 18 June 1817,
exactly two years after the Battle of Waterloo. According to *The
Gentleman's Magazine* of June 1817, the bridge was bedecked by
the national flags of Russia, the Netherlands and Prussia, as well
as of Britain, representing the allies who combined to defeat
Napoleon. A party of Horse Guards, many of them riding the
same horses as at the Battle of Waterloo, provided a guard of
honour. As the Prince and his party arrived in their barges from

Westminster, there was a 202-gun salute, representing the number of French cannon captured at Waterloo. *The Times* noted gloatingly on the following day that Napoleon had given two new bridges over the Seine the names of Jena and Austerlitz, 'where he had gained decisive victories. But these bridges, however elegant and convenient, are but trifles in civil architecture and engineering compared with that which was opened yesterday.' The Prince Regent was so impressed that he wanted to confer a knighthood on John Rennie on the spot, but, with typical Scottish self-effacement, Rennie refused. Later, he wrote to a friend: 'I had a hard business to escape a knighthood at the opening.'

The first people to cross the bridge and pay the one-penny tolls were the Prince Regent and his party. Iron turnstiles were installed at the four Doric-style toll lodges. These let just one pedestrian through at a time and were connected to machinery in the toll-gates to record the number of bridge users. This system was far in advance of the traditional manual collection system at other toll bridges. However, the large income expected from the tolls never materialised because Blackfriars and Westminster bridges were both free of tolls and people were willing to walk further to avoid paying to cross Waterloo Bridge. By 1840, the situation was so acute that the £100 shares in the Waterloo Bridge Company were worth just £1, since they provided no income after expenses and the paying off of loans.

The *Sunday Monitor* of 6 November 1825 recorded a clever attempt by two sweep boys to pay only one toll between them. At the time, policing was in the hands of the Bow Street Runners, a small body of thief-catchers set up in the eighteenth century by the novelist Henry Fielding, who was the magistrate at Bow Street Court. The name 'Runners' recognised the fact that they were much more effective than the elderly watchmen who preceded them and who were easily outpaced by criminals. Mr Skillern, one of the Bow Street Runners, noticed one of the boys climbing into their bag of soot and being carried to the turnstile, where the other boy paid his single penny toll. Skillern informed the gatekeeper and they both caught up with the boy before he

The old Waterloo Bridge and Hungerford Suspension Bridge
before the building of the Embankment

reached the south bank gate. A crowd gathered while the sack was opened and a sweep boy emerged blackened from head to toe with soot. The crowd sided with the boys and soon several shillings were raised. The money was given to the boys, who thereby made a profit despite having to pay the second toll. The bridge was eventually made toll-free after the Metropolitan Board of Works bought it for £474,000 in 1877. This valued the bridge at a little over half what it had originally cost.

In 1845, Brunel's Hungerford Suspension Bridge was constructed upstream of Waterloo Bridge to provide an extra crossing in the area for pedestrians. As shown in the illustration above, the south bank, dominated by the Shot Tower, was then very industrialised, while the embankment had not yet been constructed on the north bank. Here, Somerset House and the other classical buildings abutted directly onto the river and were approached by watergates. Towards the end of the nineteenth century, the area to the west of Waterloo Bridge, previously occupied by the old Savoy Palace, was completely transformed, apart from the Savoy Chapel, which still stands today as the only remaining building of the old palace. In 1881, Richard D'Oyly Carte bought the site for the construction of a theatre following the success of Gilbert and Sullivan's light operas such as *HMS Pinafore* and *The Pirates of Penzance*. Then, in 1889, he opened the Savoy Hotel, which was the first hotel in London to provide bedrooms with private bathrooms. The cost of a double room with bath was 12 shillings per night.

Also in the 1880s, a low red-brick building was constructed between the Savoy Hotel and Waterloo Bridge for the Institute of Electrical Engineers. This was to become the site of one of the most significant events in the history of broadcasting. After the appointment in 1922 of John Reith as the first general manager, the BBC began broadcasting from No. 2 Savoy Hill with the words, 'This is London.' Transmissions continued in the shadow of Rennie's bridge until the opening of the new BBC headquarters at Langham Place in 1932. On 14 May 1932, according to *What's On in London* of 12 January 1973, the last programme from No. 2 Savoy Hill ended with the words, 'This is the end of Savoy Hill,' uttered by the nightwatchman, Oliver, as he clanged shut the entrance doors of the building.

Although Waterloo Bridge proved a financial disaster, it continued to excite the admiration of the public and inspired many paintings. John Constable witnessed the opening ceremony in 1817 and took 15 years over his great painting of the scene, with the pageantry of boats of all shapes and sizes crowding the river and almost eclipsing the view of Rennie's bridge. The painting was first exhibited at the Royal Academy in Somerset House in 1832, and today hangs in Tate Britain. Claude Monet could see Waterloo Bridge looking eastwards from his window at the Savoy Hotel. He was struck by the contrast of Rennie's curved stone arches with the straight-line silhouette of the steel girders of Charing Cross Railway Bridge, which he could see to the west. Monet painted no fewer than 40 canvases of Waterloo Bridge, exhibiting his characteristic skill in depicting the effects of changing light, especially over the smoking factory chimneys on the south bank in the background.

Waterloo Bridge had one of the worst reputations for suicides and crime among London's Thames bridges. John Timbs states that an average of 40 people a year attempted to kill themselves by jumping into the river.[31] Mark Searle describes how five well-dressed gentlemen paid their tolls and, having passed through the gates, jumped into the river without any explanation, in an apparent act of mass suicide.[32] For this and other reasons, Waterloo Bridge was not universally popular among local tradesmen. In an undated letter, a copy of which is held in Westminster City Archives, W.C. Day complains to the Chief Commissioner of Police that

> vagabonds, costermongers and itinerants of every description injure the trade of local businesses. At 2 p.m. the penny omnibuses plying over Waterloo Bridge convey freights of transpontine prostitutes to the corner of Wellington Street, from where they take their allotted beats . . . Numerous brothels flourish under the title of coffee houses and it is utterly out of the question that any respectable female could stop to inspect a shop window.

Waterloo Bridge's reputation for prostitution seems to have endured well into the twentieth century, as implied in the film *Waterloo Bridge*. The romantic story concerns the young ballet dancer Myra and an aristocratic British army officer, Roy. They meet on Waterloo Bridge during an air raid in the First World War, fall in love and are soon engaged to marry. When Roy is called to the front, the marriage has to be postponed. Myra loses her job for missing a performance when she goes to Waterloo Station to say goodbye to Roy. She falls into poverty, and when she reads that Roy has been killed, she reluctantly decides to earn her living as a prostitute. She takes up her stand on Waterloo Bridge as the returning soldiers flood out of the station at the end of the war and meets Roy, who had mistakenly been listed as dead. Their reunion is blighted by her feeling that she is no longer fit to marry him, and she runs away to commit suicide under the wheels of an army truck on a foggy night on Waterloo Bridge. *Waterloo Bridge* was filmed twice. The first film starred Mae Clarke and Douglass Montgomery. It was made in 1931 when Rennie's bridge still stood. The second film starred

Vivien Leigh and Robert Taylor and was made in 1940, just after the start of the Second World War. An epilogue was added showing the scene on the eve of the Second World War when Roy, by now an aging general with grey hair, revisits Waterloo Bridge and recalls his lost love. By this time, Rennie's bridge no longer existed and a new bridge was under construction. A mock-up of the old bridge was used for the film.

Old Waterloo Bridge did not die without a fight. The first sign of problems occurred in 1833, when dredging of the river-bed exposed the pier foundations. Rubble, mainly consisting of Kentish ragstone, was deposited to fill in the gaps around the piers. The Bridge Company records refer to several occasions when piers needed protection by depositing rubble or surrounding the concrete aprons by timber-sheet piling. In 1867, the northern river-pier was made completely safe, as it was incorporated into the Victoria Embankment. In May 1924, one of the other piers sank. After over 100 years of river scour and irresponsible dredging, and by now having to carry new forms of motorised transport when it had originally been designed for the age of horse-drawn vehicles, Waterloo Bridge had to be closed for repairs. Sir William Arrol & Co. constructed a temporary iron-girder bridge, which was used until the bridge could be reopened after the river-pier foundations had been strengthened. After the completion of the repairs, the temporary bridge was left standing in case the old bridge should fail again. This was to prove a wise decision.

The battle now raged between the pragmatists who wanted to build a new bridge fit for modern traffic conditions and the conservatives who wanted to preserve old Waterloo Bridge for historical and cultural reasons. The former were headed by London County Council, whose engineers insisted that the old bridge was beyond repair and should be completely replaced. The latter consisted of a group of societies, including the Royal Academy and the Royal Fine Art Commission, which commissioned a report from some eminent engineers to show that old Waterloo Bridge could be strengthened and widened for £1,295,000, which was the estimated cost of a new bridge. In

1926, in an attempt to resolve the impasse, the Government set up the Royal Commission on Cross-river Traffic, which looked first at the problem of Waterloo Bridge but also considered the whole question of the adequacy of London's Thames bridges for current and future traffic volumes.

Having heard much conflicting evidence, the Commission stated:

> We realised from the outset that a controversy to which so much publicity had been given might have tended to harden the attitudes of those ranged on either side, and as we proceeded with our enquiry we found little indication of any approach to an agreement between them.[33]

The final conclusion was that Waterloo Bridge should be strengthened by rebuilding some of the piers, and that the roadway should be widened to 35 feet in order to carry four lanes of traffic. It was also recommended that a new combined road and rail bridge should be constructed at Charing Cross. As described in Chapter 9, the Charing Cross scheme never materialised; but the Government tried to progress the Royal Commission's recommendation to widen Waterloo Bridge.

Unfortunately for the old bridge, politics intervened. In 1934, Labour gained control of the LCC under the leadership of Herbert Morrison, who was not impressed by any of the aesthetic or historical arguments put forward by the Conservative government and its artistic supporters in favour of preserving Rennie's world-renowned structure. Morrison announced his decision to demolish the old bridge by personally breaking off the first stone on 21 June 1934 as he uttered his defiance of the cultural elite who opposed him with the words:

> There is no absolute standard of beauty in art. Beauty is a matter for individual decision and I am not going to delegate to the Royal Institute of British Architects or anybody else my right to decide what is beautiful.[34]

The Government initially showed its displeasure by refusing any financial contribution towards the rebuilding project, although by 1936 tempers had calmed down and a grant of £300,000 was voted through Parliament. By then, Morrison himself had shown some regret for his act of cultural vandalism. *The Times* of 9 July 1936 announced that Mr H.B. Amos, who had opened a subscription list for a memorial to John Rennie, had received a note from Morrison saying that

> [when] he struck the first blow that started the operations, he had a sympathetic and respectful thought for Rennie and hoped he would forgive him. In the circumstances he thought that he, at any rate, ought to make a small contribution.

After Morrison had ceremonially removed the first stone, Sir William Arrol was awarded the contract to remove Rennie's bridge for £331,000. Fortunately, the old temporary bridge built in 1924 still stood, so traffic could continue to cross the river during the demolition process. *The Builder* of 20 March 1936 commented on the widespread regret at the end of the old Waterloo Bridge and related that

> an engineer and a lay friend were walking along the embankment and the latter remarked, 'How sad it is to see one of Wren's masterpieces being ruthlessly torn down.'
>
> 'Yes,' said the engineer, and unwilling to expose his friend's ignorance, added, 'but we engineers in our affectionate way call him Rennie.'

During the demolition project, a workman found the foundation stone of the old bridge, as well as the glass container with gold, silver and bronze coins dating from the end of the eighteenth century which had been deposited underneath at the stone-laying ceremony in 1811. On 4 May 1939, the foundation stone of the new bridge was laid using part of the old bridge's foundation stone. A copper cylinder of contemporary coins, stamps and daily papers was deposited under the stone.

Sir Giles Gilbert Scott (1880–1960), designer of Battersea and Bankside power stations, was appointed as architect for the new bridge. He worked with the engineering firm Rendel, Palmer and Triton to design the first reinforced-concrete bridge to cross the Thames in central London. The bridge has a 58-foot-wide roadway and two 11-foot-wide footways for pedestrians. Its total length is 1,200 feet, consisting of five 240-foot spans. Although at first sight the spans look like impossibly flat arches, they are in fact steel box-girders which form the basis of the reinforced-concrete structure. The northern span lies half over the river and half over the Victoria Embankment. The next two spans cross the river where the water is deep and are normally used for boats to pass through, while the two southern spans cross shallow water. The sides of the bridge are faced with Portland stone, which was chosen because the underlying concrete does not wear so well and was considered unsuitable for the central location with impressive stone buildings such as Somerset House in the vicinity. The concrete structures that line the south bank here today did not exist at the time. The river-piers which support the superstructure are not visible because it was decided to surround them by a granite and Portland stone-faced shell. This innovatory design allows the piers protection from collisions with shipping as well as providing permanent cofferdams in case repairs should be required.

Although the design of the new bridge could never rival Rennie's masterpiece from an aesthetic point of view, its simple if severe elegance was generally well received. Even the president of the Royal Academy, who had fought so hard for the preservation of the old bridge, gave it his approval. The new bridge's engineering design received universal praise and was widely admired in the profession. As recorded in the 1943 Institution of Civil Engineers *Proceedings*, Peter Lind, whose company won the construction contract, commented that he did not think that an ounce of steel or concrete had been put in which could not be held to serve a useful purpose.

Somerset House glimpsed through the southern arch of Waterloo Bridge

The construction project was delayed considerably both by two strikes which occurred in 1937 and 1939, and by labour shortages and enemy action during the course of the Second World War, which started when the superstructure was still resting on temporary timber staging. There was great concern that the bridge would collapse in war conditions and block the whole river. Fortunately, despite 20 air-raid incidents which caused damage both to the old temporary bridge and to the new bridge itself, work progressed well enough for the new bridge to be partially opened for two lanes of traffic in August 1942, while the footways were opened in December. According to figures presented by E.J. Buckton at the June 1943 meeting of the Institution of Civil Engineers, the numbers of men working on the project between 1937 and 1942 varied from 100 to 500. In fact, as a result of the severe shortage of manpower on the home front during the Second World War, many of the workers were women. For this reason some people still refer to Waterloo Bridge as the 'Ladies' Bridge'.

Unlike the 1817 opening of Rennie's bridge, which

commemorated Wellington's famous victory over Napoleon, the partial opening of the new bridge was conducted without ceremony because of wartime conditions. The *Evening Standard* of 11 August 1942 reported that

> No Cabinet Minister was present to give it a blessing, no LCC chief to represent the ratepayers who paid for it and only a small crowd, among them bicyclists, a horse-drawn hackney cab, several motor cars and a bus laden with passengers.

At precisely 10 a.m., the foreman walked across the bridge to remove the red flags and a race began to see who could cross over first. The race was won by a 16-year-old schoolboy on a bicycle, who got a flying start and so reached the winning post before the faster motorised vehicles could catch him.

In 1943, the temporary bridge which had served its purpose well for nearly 20 years was at last demolished. The huge steel girders were then shipped to Belgium and performed their final duty during the Allied advance on Germany near the end of the Second World War. In March 1945, the US First Army was advancing to cross the River Rhine at Remagen when the Germans tried to impede progress by setting off high explosives on the bridge, which had originally been built there by General Ludendorff during the First World War. The superstructure of the bridge was lifted up in the air, but, amazingly, it dropped back down onto its supports, thus allowing the Americans to cross the river and set up defensive positions on the other side. Unfortunately, the bridge had been severely weakened, and on 17 March it collapsed into the Rhine while engineers were attempting to strengthen it. Ten men were killed and eighteen reported missing. The temporary Waterloo Bridge was then re-erected in one week under enemy fire, and so the Allied advance was hardly interrupted. Having served its purpose, the temporary bridge was finally demolished. The Ludendorff Bridge itself was not rebuilt, although its bank-side towers remain standing and have been turned into a Museum of Peace.

Meanwhile, in London, the new Waterloo Bridge was opened in November 1944 for all six lanes of traffic, again without ceremony. After the end of the war, the official opening was performed on 10 December 1945 by Herbert Morrison, who had started the destruction of Rennie's bridge 11 years before. Even after its demolition, remains from the old Waterloo Bridge can still be found in many parts of the country today. The river-piers of the new bridge are faced below water level with granite from the old bridge. Underneath the present bridge, on the Victoria Embankment, is a platform which was built over the foundations of one of the piers of Rennie's bridge, with the original twin Doric columns on either side. On the river side of the platform is a bronze model of the old bridge and a replica of the 1811 foundation stone plaque. These were erected here in 1974. Sadly, few people apart from the homeless visit this rather gloomy spot under the shadow of Waterloo Bridge. Two of the original lamp-posts now adorn Rennie's earlier bridge at Kelso in the Scottish Borders.

Sixty thousand tons of stone from the old bridge were taken away for storage at Harmondsworth, Middlesex. According to the *South London Press* of 10 February 1961, much of this was eventually bought by the contractor Michael Pierce, who planned to crush it for use as under-surfacing for the new M4 motorway. When he found out its origin, he asked Sotheby's to sell it off piecemeal as a piece of history, and it was used in many a gravestone, fireplace and sundial throughout the land. Appropriately, the memorial to John Rennie in his birthplace at East Linton contains a bird bath standing on a pedestal made from one of the balusters of his Waterloo Bridge. Sadly, the old balustrading and Doric columns which were originally incorporated into the southern abutment of the new bridge were removed during the 1960s redevelopment of the South Bank cultural centre, when the Hayward Gallery, Queen Elizabeth Hall and National Film Theatre were constructed.

The only major development on the northern side of the new bridge involved the subway to Kingsway. In 1908, the LCC had constructed a tram subway to link Kingsway to the Victoria

Embankment as part of the major redevelopment of Aldwych. When Waterloo Bridge was reconstructed, the subway had to be diverted to align with the new wider bridge in order to allow the trams to emerge from under the centre of the bridge and turn on to the embankment. The last tram passed through in 1952, and the subway was used as a storage area since nobody could agree what to do with it. In 1959, it was finally agreed to use the subway as a basis for a road link from Waterloo Bridge to Kingsway for northbound traffic. The new ramp leading down to the subway had to start clear of the bridge structure and therefore required a steep gradient of 1 in 12 to allow it to pass under the Strand sewer. This meant that restrictions were placed on bicycles and vehicles over 12 feet high.

Today, Waterloo Bridge is one of London's busiest bridges both for traffic and pedestrians. On 7 September 1978, the evening rush hour on Waterloo Bridge proved the ideal spot for a Cold War political assassination with all the elements of a James Bond plot. The victim was Georgi Markov, who had become one of Bulgaria's most respected writers under the

The City skyline viewed behind Waterloo Bridge's elegant flat white arches

Communist regime of Todor Zhivkov. Increasingly disillusioned by the lack of freedom and the personality cult that had developed around his country's ruler, Markov left Bulgaria clandestinely in 1969 and never returned. His broadcasts for Radio Free Europe and the BBC enraged the Bulgarian leadership to such an extent that an agent was sent over to England to warn him that he would be eliminated if he did not cease his criticisms. Markov refused to be intimidated, and on the fateful evening of 7 September 1978, on Waterloo Bridge, he felt a sudden stab of pain in his thigh. Turning round, he saw a swarthy man who bent down to pick up an umbrella and muttered 'I'm sorry' in a foreign accent before hailing a taxi.

On returning home, Markov thought no more about the incident, but within three days he was dead. The post-mortem established that a precision-made ricin pellet embedded in his leg was the cause of death and that it must have been fired from an umbrella with a built-in compressed air cylinder, which had been developed by the KGB. This was the same weapon used a few weeks before in an unsuccessful attack on another Bulgarian defector, Vladimir Kostov, in Paris. Although suspicion fell heavily on the Durzhavna Sigurnost (DS), the Bulgarian secret service, the actual perpetrator could not be identified until the daily newspaper *Dnevnik* published leaks from its files in May 2005. These revealed that the killer was Francesco Giullino, a Danish citizen of Italian origin, who had been caught smuggling drugs into Bulgaria and in return for avoiding punishment had become a DS spy. The whereabouts of Giullino and his umbrella are still unknown, although when the Communists were ousted from Bulgaria in 1989, a whole stack of the special KGB umbrellas was found in the interior ministry. Markov, who had settled in England, is buried in a Dorset country churchyard far removed from the intrigues and violence of the Cold War. It is to be hoped that his death signalled the end of Waterloo Bridge's long connection with battles and wars.

CHAPTER 11

Blackfriars

At Blackfriars, there is a cluster of three bridges, although the middle one now consists merely of headless columns, as the railway crossing they supported was demolished in 1985. The upstream bridge is Blackfriars Bridge, opened in 1869 to replace the original Blackfriars Bridge, which had been completed almost exactly 100 years before. Just downstream from the headless columns of the former London, Chatham and Dover Railway Bridge is another railway bridge which also belonged to the LCDR. This was completed in 1885 and is the only bridge to carry railway services across the Thames from south to north through the centre of London.

Blackfriars Bridge
The name of Blackfriars Bridge commemorates the former Blackfriars monastery which was sited at the north-east end of the bridge until it was dissolved by Henry VIII during the Reformation. The origins of the monastery go back to 1276, when a community of Dominicans, otherwise known as Black or Preaching Friars, was given land near Ludgate on the site of Montfichet and Baynards castles, which had been built during the reign of William the Conqueror. Great monastic buildings were constructed in place of the demolished castles. Montfichet

Castle was never rebuilt and is virtually forgotten, while Baynards Castle was rebuilt downstream of Blackfriars monastery, only to be destroyed in the Great Fire of 1666. It is still remembered in the name of Baynard House, where BT has a rather forbidding concrete office block.

Blackfriars monastery had a distinguished history. On 17 January 1382, William Courtenay, Archbishop of Canterbury, held a council of bishops and lawyers to examine and condemn 24 articles drawn from the teachings of John Wycliffe. This became known as the 'Earthquake Council' because a great earthquake shook the city that day. Wycliffe famously stated that this was a sign of God's judgement on those who had condemned his teachings. In 1522, Charles V, the Holy Roman Emperor, stayed at Blackfriars as the guest of Henry VIII, and in 1529 Henry VIII attended the court here when the papal legate heard the case for his divorce from Catherine of Aragon. Blackfriars was dissolved in 1538. Most of the buildings were demolished at that time, and the remainder were destroyed in the Great Fire of 1666.

All that remains of the monastery in the area is a fragment of wall in Ireland Yard and the names of the bridge, the station, some streets and the Black Friar pub in Queen Victoria Street. On the outside of the pub is a mural depicting some portly friars catching fish in the River Fleet, which used to flow past the western side of the monastery. The twin sources of the Fleet in Hampstead and Highgate still exist, but the lower course of the river, which flowed through King's Cross and down to the Thames at Blackfriars, has long since vanished underground. When crossing Blackfriars Bridge today, it is hard to imagine the time when a wide river flowed into the Thames at this location. The River Fleet had formed the western boundary of the ancient Roman city of Londinium. According to John Stow, the Fleet had been 'of such breadth and depth, that ten or twelve ships' navies at once, with merchandise, were wont to come to the bridge of Fleet'.[35] The river was subject to pollution from many sources, including offal from Smithfield meat market, private and chargeable latrines, and rubbish of all sorts and sizes. Stow records several attempts to clean up the river, the latest in 1589, but says that at the time of writing in 1598 it was 'worse cloyed and choken than ever it was before'.

Following the Great Fire, Sir Christopher Wren again cleaned up and canalised the lower course of the River Fleet as far as Holborn. However, few ships used it, and soon the river became so silted up with pollution again that it was referred to as the 'Fleet Ditch'. Jonathan Swift described the state of the Fleet in his poem of 1710:

> Now from all parts the swelling kennels flow,
> And bear their trophies with them as they go:
> Filth of all hues and odours seem to tell
> What street they sailed from, by their sight and smell.
> [. . .]
> Sweepings from butcher's stalls, dung, guts and blood,
> Drown'd puppies, stinking sprats, all drenched in mud,
> Dead cats, and turnip tops, come tumbling down the
> flood.[36]

The state of the Fleet clearly provided cause for amusement as well as disgust, as in the story told in *The Gentleman's Magazine* of 1844 about a boar which had disappeared from a butcher's stall in Smithfield but had turned up five months later at the mouth of the river, which by then had been covered over. The result of the boar's sojourn in the sewer was such considerable fattening that it was improved in price from ten shillings to two guineas.

The stretch of the river between Holborn and Fleet Street was covered over in 1733 for the establishment of the Fleet Market. There was little local opposition when the City Corporation submitted a Bill to Parliament in 1756 to cover over the final stretch below Fleet Street with a view to constructing a bridge over the Thames. As we have seen, the City Corporation had previously strongly opposed any attempt to construct a bridge across the Thames in London. However, now that Westminster Bridge was completed, the fear was that the centre of business might shift away from the City, and this caused a radical rethink about strategic river crossings. According to the Bill, there were many benefits to constructing a bridge between Westminster and London bridges. It was argued that such a bridge would provide a crossing in what was then the centre of London, between Westminster in the west and London Bridge in the east; it would combat the grand schemes for the area around Westminster Bridge that threatened to take business away from the City; it would result in new streets and developments on the south bank around St George's Fields and therefore increase the value of the City's estates in that area; it would provide an alternative crossing in the likely event that London Bridge closed for repairs or widening; and it could outdo Westminster Bridge in grandeur and magnificence. George Dance (1700–86), the distinguished City Architect, produced a plan for a bridge between what was by then known as Fleet Ditch and the opposite side of the river for an estimated cost of £185,000. A report delivered to the Common Council of the City Corporation on 26 September 1754 recommended that the bridge costs should be met by Parliament because it would be of benefit to the nation. It also recommended that the houses on London Bridge should be removed.

In 1756, an Act was passed 'for building a Bridge across the River of Thames from Blackfriars within the City of London, to the opposite side in the County of Surrey'. Authority was given to raise up to £160,000, but the proposal that costs should be met at the national expense was rejected, presumably because the main reason behind the proposal was to benefit the City in competition with Westminster. The money raised was to be paid off by tolls which should cease as soon as the debt was cleared. One of the many stipulations in the Act prohibited the building of any houses on the bridge, except for tollhouses and toll-gates. Authority was given to provide access roads and to 'fill up the channel of Bridewell Dock up to Fleet Bridge and to build and maintain a sewer to the Thames under the new road'. The Act also allowed the City Corporation to remove the houses from London Bridge. The only people who had opposed the Act were the City's old allies, the watermen, who put forward the usual arguments about danger to navigation and silting up of the river that the City authorities themselves had used in opposing all previous attempts to construct a bridge across the river in London. Since the watermen ran a Sunday ferry at Blackfriars, the Act did provide compensation to them for loss of business.

The City Corporation now set up a Blackfriars Bridge Committee to oversee the execution of the works specified in the Act. Its first job was to select the bridge design. As with Westminster Bridge, much controversy surrounded the selection process and, again as with Westminster, there was a surprise winner. In 1759, the Blackfriars Bridge Committee chose a shortlist of eight, including the design of the City Architect, George Dance. The other competitors included William Chambers, John Smeaton, John Gwynn and a little-known young Scottish architect, Robert Mylne, who had no real experience but had won a first-prize medal for architecture during a five-year period in Rome, where he had been influenced by Piranesi. Smeaton had just completed the famous Eddystone Lighthouse and, with his experience of building in water, was considered the favourite. Chambers and Dance were the outstanding architects of their time and were also considered strong contenders.

However, in 1760 a paper was published, under the anonymous authorship of Publicus, which analysed the shortlisted designs and attacked them all apart from Mylne's.[37] Publicus claimed that Chambers, although a distinguished architect, had produced a design too imitative of Roman precedents and unsuitable for England. Dance had grown old in the service of the City and his design was now old fashioned. Publicus had expected Smeaton to produce 'a most complete piece of mechanism. This, however, I don't find quite so clear, but on the contrary it is from ocular demonstration the weakest of all the designs given in.' Mylne, on the other hand, had produced a design 'magnificent, yet simple; light, yet not too slender; and modest, without bordering on rusticity'. Praise was especially given for his concept of niches for statues of admirals to celebrate England's domination of the sea, although after Mylne finally won the competition these statues were not erected for reasons of cost.

Robert Mylne (1734–1811)

Robert Mylne was the eldest son of Thomas Mylne, surveyor of the City of Edinburgh, who favoured the classical style of architecture which was prevalent at the time. In 1754, he went to Rome to complete his architectural education. There, he won a first-prize medal for architecture at the Academy of St Luke, and his early interest in classical architecture was reinforced by his friendship with Piranesi. In 1759, he returned to England just in time to enter the competition for the design of Blackfriars Bridge. Although his training was in architecture, he had obtained considerable skill in engineering. At that time, these professions were not as clearly defined as they are today.

Following his success at Blackfriars, he was appointed surveyor of both Canterbury and St Paul's cathedrals. He obtained many commissions for bridges and villas throughout England and Scotland, which involved much travel even during his time as manager of the Blackfriars Bridge project. His other major appointment was as engineer of the New River Company, which had been set up in the seventeenth century to construct a channel to bring water from rivers in Hertfordshire to supply much of London's water from its reservoir in Clerkenwell.

There had been much resentment when Mylne, as a young man from Scotland, had defeated his more famous English rivals in the Blackfriars Bridge competition. However, by the end of his life he could move in the highest circles, having been elected a Fellow of the Royal Society in 1767. He was buried near to Sir Christopher Wren's tomb in St Paul's Cathedral.

Needless to say, Publicus's pamphlet aroused fury among Mylne's rivals. Smeaton wrote his own rebuttal of Publicus's criticisms and challenged the author to admit that Publicus was in fact Mylne himself. Publicus replied, repeating his arguments against Smeaton. He refused to reveal his identity and, addressing the Committee directly, claimed that his only motive was to 'save my fellow citizens and you, our trustees'. At this stage, a surprising intervention came from Dr Samuel Johnson, who was the most distinguished figure in London's intellectual society at the time but was not known for his knowledge of engineering, as was admitted by his normally appreciative biographer James Boswell. He wrote a letter to the *Daily Gazetteer* criticising Mylne's use of elliptical arches, which Johnson claimed would be much weaker than the semicircular arches proposed by Johnson's friend John Gwynn. Unfortunately, Johnson had failed to understand that Mylne, by use of ingenious counter-arches built into the stonework of the bridge, had ensured that his elliptical arches would be as strong as semicircular ones. Moreover, the expert whose opinions Johnson quoted to support his arguments wrote that he was not in fact opposed to elliptical arches.

The ever-irascible Johnson refused to give up and published two more letters with rather more personal attacks on Mylne, but he had clearly lost the argument. Johnson was accused of attacking Mylne because of his well-known prejudice against the Scots. Boswell, however, claims that Johnson's real motive was to support his friend Gwynn.[38] Moreover, he adds that: 'So far was he from having any illiberal antipathy to Mr Mylne that he afterwards lived with that gentleman upon very agreeable terms of acquaintance, and dined with him at his house.'

Having selected Mylne's design, the Blackfriars Bridge Committee decided to appoint him as surveyor and engineer, effectively putting him in charge of the project. The bridge was 995 feet long and 45 feet wide, supported on nine semi-elliptical arches. Between the arches were double Ionic columns which supported small projecting recesses against the face of each pier, but with no statues. The first stone was laid by the Lord Mayor, Sir Thomas Chitty, on 31 October 1760. The ceremony took place in the middle of the Seven Years War, when Britain was allied with Frederick the Great of Prussia against France, Spain and Russia. General Wolfe had just won the battle for Quebec, losing his life in the process, and patriotic fervour was at its

Robert Mylne's Blackfriars Bridge of 1769 as painted by William Marlow

height. The Lord Mayor read a glowing inscription praising the Prime Minister, William Pitt the Elder, for 'augmenting and securing the British Empire' and declaring that: 'The Citizens of London have unanimously voted this Bridge to be inscribed with the Name of William Pitt.'[39] The inscription was deposited under the foundation stone, together with contemporary coins, a copy of that day's *Times* and the medal Mylne had won in Rome. According to Mylne's grandson, as recorded in *The Times* of 5 July 1865: 'In the enthusiasm of the moment the architect took the medal from his neck and threw it into the cavity of the stone.' Well before the bridge was completed Pitt had lost his popularity with the citizens of London because of the great expense of the war, and it was opened as Blackfriars Bridge in 1769.

As with Westminster, then, Blackfriars Bridge was constructed partly during time of war. The war was one of many reasons given by Mylne for the nine-year length of the project and the escalating costs. The final cost was £232,000, which was more than the £160,000 authorised in the first Blackfriars Bridge Act, but, as Mylne pointed out to the Committee, much less than the

£400,000 for Westminster Bridge. Extra money amounting to £16,200 was voted by the Common Council to come from the fund of sheriffs' fines. These fines were paid by sheriffs, who were senior aldermen due to be appointed Lord Mayor of London but who preferred to buy themselves out of this honour because it would involve considerably more personal expense than the fine.

The Bridge House Estates (BHE), which had been set up to finance London Bridge, was under pressure at the time because of the removal of the houses on London Bridge and the resulting loss of income. The City Corporation was also involved in the expenses of removing all the Roman city gates and rebuilding Newgate Prison. Consequently, the rest of the money had to come from extending the tolls from their expected end date of 1770 until 1785, and the Sunday tolls until 1811. This was authorised by an Act of Parliament in 1769. Since pedestrian tolls were just one penny, a huge amount of coinage was collected. The Blackfriars Bridge Committee had an agreement with a Mr Cordy that he would pay £10 for every 100 lb of copper. As recorded in the minutes of a meeting of the Committee on 27 June 1775, Cordy complained in a letter that so much bad money had been collected that he made a loss on the deal.

Mylne's diaries describe the many problems he met with during the construction project. In the winter of 1762–3, the Thames froze over. The Frost Fair held on the frozen river provided a variety of public entertainments but stopped any work on the bridge. Like Labelye at Westminster, Mylne used caissons for the construction of the pier foundations. However, having learned from the disaster that befell one of the Westminster piers, Mylne had wooden piles driven deep into the river-bed to give firm support before building the piers. Even so, the caissons seem to have caused severe difficulties, with four men working around the clock on pumping them dry. Attempts to float the caissons frequently failed, and on one occasion four horses, sixty men and nine pumps proved inadequate for the job. The Fleet Ditch must have added to the difficulties faced by the workmen. According to a contemporary witness:

The large sewers that empty themselves in the neighbourhood occasion a constant accumulation of sand, mud and rubbish which destroys a great part of the navigation at low water. The mud and filth thus accumulated is extremely offensive in summer and often dangerous to the health of the neighbouring inhabitants.[40]

Surprisingly, Mylne himself records only one accident during the whole of the project, when 'one man's arms were a little hurt when a great part of the timbers supporting the arches broke'.[41] There would certainly have been many more serious accidents and fatalities on so dangerous a project given the lack of modern equipment and safety standards. This is hinted at in the 1784 report of the Blackfriars Bridge Committee, which mentions 'the many and great unavoidable accidents, commonly attendant on all great undertakings'.

Detailed information on such matters was lost when in 1780 the records of the Blackfriars Bridge Committee were destroyed in the Gordon Riots. The Gordon Riots were instigated under the banner of 'No Popery!' by Lord George Gordon, who objected to the repeal of anti-Roman Catholic legislation. He assembled a crowd of 30,000 in Southwark, but soon lost control. The crowd quickly forgot about the Catholics and became a mob, rampaging through the streets of London, destroying many of the prisons, including the newly rebuilt Newgate Prison, and freeing the prisoners. They also vented their fury on the Blackfriars Bridge tollhouses. The Blackfriars Bridge Committee minutes of 12 June 1780 record that on the night of 7 June, known as 'Black Wednesday', a great mob attacked Blackfriars Bridge and fired the tollhouses. They carried off the money chests, which were found empty of all but half a guinea and four sixpences after the mob had been dispersed by the military. Early records of the Committee were also destroyed. The general riots finally died down when the troops were called in and many of the rioters had had too much to drink. Gordon himself was arrested. Ironically, he died soon after in Newgate Prison, after it had been reconstructed yet again.

Meanwhile, Mylne continued to be the surveyor of Blackfriars Bridge and was responsible for rebuilding the toll-gates. He also built himself a substantial house in the now more fashionable area to the north of the bridge, where, as recorded above, he entertained Dr Johnson to dinner. It is surprising that he kept the position through all those years, because he had been involved in a protracted dispute with the Common Council about payment for his work on the construction of the bridge. The initial contract had stipulated that he should be paid a salary of £350, but there was a general understanding that people in his position should receive 5 per cent of the construction cost and 1 per cent of the purchase cost. Since there was nothing about this in the written contract, Mylne presented a petition claiming what most people agreed was his due. In 1771, a court of the Common Council examined Mylne and asked him if he thought he was entitled to the full payment as a right. Mylne answered that he did claim it as a right. It seems that Mylne's answer was considered arrogant, and the petition was turned down. Mylne then resubmitted his petition in 1774 to the Blackfriars Bridge Committee, which agreed he should be paid the extra money but could not authorise the payment without the approval of the Common Council. Eventually, in 1776, Mylne received the exact amount he demanded, minus his annual salary, as evidenced in his diary entry for payment of £4,209 16s. ¾d.

One of the reasons for the length of the construction project and the cost overruns was the difficulty in negotiating property purchases for the approach roads. On the northern side, New Bridge Street and Chatham Place were constructed, and these were soon lined with impressive buildings, including Mylne's own house. Underneath the buildings ran the Fleet Ditch, which was now a 17-foot-wide sewer diverted to empty into the Thames clear of the bridge. In 1783, an inspection established that the roof of the sewer was in danger of collapsing, which would result in many buildings falling into it. Expensive repairs were needed to ensure the safety of the buildings. At the same time, embankments were constructed from Blackfriars to Temple to overcome the problem of vessels accessing the quays and wharves, which were silted up at low water.

The southern approaches proved even more problematic because of the many and complex land rights in St George's Fields. The original idea was to construct roads to London Bridge in the east and to Westminster Bridge in the west, but this proved too complicated. Eventually, a compromise was reached with the trustees of the Surrey Turnpike to construct a road to St George's Circus, where roads from London and Westminster bridges already converged. The City funded the road, but the Turnpike Trust was made responsible for lighting, watching and repairing it from tolls collected at the toll-gate at St George's Circus. At the time, the road was known as Great Surrey Street, but it was renamed Blackfriars Bridge Road in the nineteenth century. An obelisk was erected at St George's Circus in 1771, giving the date of construction under the mayoralty of Brass Crosby, as well as mileages to the three bridges. This was removed in 1905 but has now been restored to its original position, where it is noticed by few people because of today's hectic traffic conditions.

The final report of the Blackfriars Bridge Committee was produced in 1786. This announced that all the work had been completed and all the capital paid off. There was concern that, now the tolls had ceased, the small residual annual income of £175 would be insufficient for maintenance. Already in 1790 it was reported that the road surface on the bridge was wearing away due to increased traffic following the removal of the tolls, and a proposal was put forward for granite paving. More severe problems occurred because of the poor quality of the Portland stone used in the bridge's construction. According to Thornbury, only the Government had use of the best Portland quarries.[42] The City Corporation, as a private enterprise, was forced to use second best.

By 1833, the deterioration was such that a new Act was passed authorising the restoration of the arches and the lowering of the crown of the bridge at a cost of £74,000. This work would lessen the architectural beauty of Mylne's bridge but was expected to prolong its life by many years. However, after the removal of Old London Bridge, the resulting increase in tidal flows caused severe damage to the foundations of the piers at Blackfriars, just as at Waterloo and Westminster bridges. By 1860, the City was so worried about the state of Blackfriars Bridge that the Common Council resolved to replace it.

This time, the bridge was to be financed from the Bridge House Estates. Proposals were submitted for a bridge of either three or five spans. The watermen and other river users favoured a three-span bridge because this would provide clearer navigation. The committee set up to examine the matter reported back to the Common Council with a recommendation for a three-span bridge. However, many members of the Common Council supported a five-span bridge, which would align with the preferred design of the proposed bridge of the LCDR to be built 100 feet downstream.

When the time came for the Blackfriars Bridge Committee to be re-elected, the supporters of the five-span bridge managed to change the membership so that the previous recommendation was revoked and invitations were sent out for a five-span design. Unsurprisingly, the design of Joseph Cubitt (1811–72), the LCDR engineer, was accepted. William Lucey, who had sponsored three petitions in favour of the three-span design, wrote a furious pamphlet summarising the arguments from his point of view and implying a degree of shady dealing:

> May I be permitted to ask the question whether the members of the Common Council have employed paid persons to get up and obtain signatures to petitions to aid them in procuring the adoption of their favourite scheme – a system which must be pronounced by all right-minded persons corrupt and nefarious?[43]

The 1863 Act authorised the removal of Mylne's failing structure and the erection 'in lieu thereof upon a deeper foundation, a new Bridge of grander width, with improved gradients, and a lesser number of arches'. Joseph Cubitt designed a bridge consisting of five wrought-iron spans with a total length of 963 feet and an unprecedented 75-foot width. The piers are ornamented with massive red polished-granite columns. The columns' capitals are carved with interlaced plants and birds, the downstream capitals representing maritime species and the upstream capitals inland creatures.

The project started in 1864, when the old bridge was demolished together with a number of nearby buildings, including Mylne's own house. Mylne himself had died in 1811. He is buried in the crypt of St Paul's near the tomb of Sir Christopher Wren. Wren's own memorial, inscribed on his tomb, reads '*Lector, si monumentum requires, circumspice*', translated as, 'Reader, if you seek his memorial, look around you.' Sadly, Mylne's great work served as his memorial for only half a century.

The foundation stone of the new bridge was laid by the Lord Mayor, Warren S. Hale, on 20 July 1865. The 2-ton granite block came from the old bridge. The inscription refers to the difference between the time of war when the old bridge had been started in 1760 and the 'profound peace in the 29th year of the reign of Queen Victoria at a moment when . . . by the adoption of Free Trade, those separate interests which divided nations have been happily bridged over'. With more modern equipment, the construction project took less than half the time required for the building of the original bridge. The *Illustrated London News* of 3 December 1864 was especially impressed by the newly invented steam cranes, erected on rails alongside the construction site, which could lift and carry the largest granite blocks with 'unexampled rapidity'.

The new Blackfriars Bridge was opened by Queen Victoria on 6 November 1869, almost exactly 100 years after the opening of the earlier bridge. Clearly the authorities were worried about the possible reception the Queen might receive after she had for so long refused to appear at any public ceremonies, including the opening of Westminster Bridge, following the death of Prince

Blackfriars Bridge of 1869 with the dome of St Paul's Cathedral

Albert. The *Illustrated London News* of 6 November 1869 printed a strong defence of the Queen's behaviour:

> If it has taken her a somewhat protracted time to recover from the effects of that crushing blow, it shall be borne in mind that Royalty, from its very position, is peculiarly isolated, and that the loss to the Queen of nearly the only companion with whom she could share the inmost thoughts, the ever-present responsibilities and the tenderest affection of her being was one which left her, more than others who have suffered a similar bereavement, solitary and alone.

On the day, according to the *Illustrated London News* of 13 November 1869, the crowds cheered Queen Victoria on her route to Blackfriars from Paddington Station, to where she had taken the train from Windsor. An alternative version, claiming that Queen Victoria and her manservant John Brown were hissed on their way to the opening ceremony, appears in *The*

London Encyclopaedia, but I have found no other evidence to support this.

The Queen's first function was to open Holborn Viaduct, which had been constructed to cross Farringdon Road over the former valley of the River Fleet. She then proceeded to Blackfriars Bridge, which she opened and then crossed in her coach. It has been said that Queen Victoria approved the bridge's red granite columns and the semicircular recesses supported by them, which were designed to look like pulpits and so be a reminder of the ancient Blackfriars monastery from which the bridge took its name. The statue of Queen Victoria by C.B. Birch on the north side of the bridge was erected in 1896. Her Diamond Jubilee was celebrated in the following year, by which time her popularity was such that the route to St Paul's Cathedral was lined by cheering crowds of over one million people.

When Cubitt's Blackfriars Bridge was completed, the Victoria Embankment had finally removed the putrescent mudbanks

The 'pulpit' capital of a Blackfriars Bridge river-pier

and the danger to health caused by the continuing outflow of sewage from the covered River Fleet. The channel of the sewer was now diverted back to its original course, so that today it comes out into the Thames underneath Blackfriars Bridge, where it can be seen at low tide from the Embankment. Bulk sewage is now taken to Beckton sewage works via Bazalgette's low-level intercepting sewer, while the outflow of the Fleet is used only as an overflow sewer in conditions of heavy rainfall.

The bridge itself was a solid enough structure, but problems arose when the London County Council presented a Bill to Parliament in 1905 with a proposal to run its trams over the river at Blackfriars and Westminster. The City opposed the Bill, but was persuaded to allow a tramway to be built at the same time as the bridge was widened. The LCC agreed terms to lease the tramway and maintain it. The Corporation of London (Blackfriars and other Bridges) Act of 1906 authorised the work to widen the bridge and install a tramway. Finance again came from the Bridge House Estates, supplemented by the LCC lease. Sir Benjamin Baker (1840–1907), who had worked with Sir John Fowler on the Forth Railway Bridge, was engaged as engineer. His design involved removing the columns and ironwork from the upstream side of the bridge and reinstalling them after the bridge had been widened from 75 feet to 105 feet. Baker was asked to manage the project but was advised by his doctor not to undertake any more business commitments because of his state of health. He died in 1907 and his partner, Basil Mott (1859–1938), took over management of the project. Sir William Arrol & Co. won the £210,000 contract, which contained bonus and penalty clauses. The project was due to be completed in 36 months, but a bonus of £20 was to be paid for each day saved and a penalty of £1 imposed for each day lost. This was to cost the City an extra £3,560 when Arrol finished the work 178 days early. Arrol had been responsible for constructing a number of outstanding river bridges, including the Tay and Forth bridges and the steelwork for Tower Bridge.

Two fatal accidents occurred during the widening project, as reported by Basil Mott to the Bridge House Estates Committee on 14 December 1911. On one occasion, a labourer fell into a barge and died a few hours later, and on the second occasion, a serious accident occurred when the wooden staging from which a caisson was being lowered collapsed, killing four men. A Board of Trade inquiry found that no blame could be attached to anyone, since unfortunately the only ones who could have confirmed what happened had lost their lives. Compensation was paid to the men's families.

The widened bridge and tramway were opened by the Lord Mayor, Sir George Truscott, on 14 September 1909, when a joint party of City Corporation and LCC dignitaries crossed the bridge in a tram driven by the Lord Mayor himself. The Lord Mayor seems to have enjoyed being in control of such a powerful machine, as the tram was seen to lurch forward at great speed. The opinions of the rest of the distinguished party were not recorded.

Blackfriars Bridge is today one of the widest and busiest of London's Thames bridges. Its history is depicted in a series of tiled mosaics underneath the south bank arches. The bridge is still notorious for a murder mystery as sinister as the poisoned umbrella murder for which Waterloo Bridge is known. On 18 June 1982, the body of a man was found hanging under Blackfriars Bridge, with bricks and stones stuffed into his pockets to weigh him down. The body was identified as that of Roberto Calvi, who was chairman of Banco Ambrosiano and who had close links with the Vatican, hence becoming known as 'God's banker'. He was also suspected of having links with the Mafia.

The initial coroner's inquiry found that Calvi had committed suicide after the collapse of his bank in the wake of news reports regarding a 400-million-pound anomaly in its accounts. He was also due to be tried for his involvement in allegedly fraudulent property deals with a Sicilian banker. However, his son Carlo campaigned to have the case reopened and in July 2003 a panel of judges in Rome issued a statement that it had no doubt that Calvi had been murdered. The judges nominated four people with alleged Mafia links as suspects. The City of London police reopened the murder case in September 2004. The trial started in October 2005 and is likely to expose the murky Mafia underworld and financial scandals possibly involving Italy's political and religious establishment. Calvi was a member of a secret Masonic lodge and it seems that Blackfriars Bridge may have had some Masonic significance. The repercussions of the trial may well put *The Da Vinci Code* in the shade.

London, Chatham and Dover Railway Bridge

The LCDR Company was described as 'one of the most gigantic frauds ever perpetrated' in the *Railway Times* of 10 December 1870. However, the *Railway Times* was a strong supporter of the rival South Eastern Railway Company, whose railway bridge over the Thames at Cannon Street was built soon after the LCDR Bridge. The cut-throat competition between the two companies resulted eventually in both becoming virtually bankrupt, and they were forced to merge in 1899. The fascinating and complex story of the LCDR and its various stations on both sides of the river has been told by Adrian Gray.[44] Here I will concentrate on the origins and fates of the two LCDR Thames railway bridges.

In 1860, the Metropolitan Extensions Act authorised the LCDR to extend its lines across the Thames at Blackfriars and Victoria. LCDR trains first crossed the Thames in 1862, to Victoria, where the company leased the western half of the station from the London, Brighton and South Coast Railway. In 1864, the LCDR opened Blackfriars Bridge Station on the south bank as its own London terminus for the lines to the south. However, this was a temporary measure, since the real aim was to be the first railway company to enter the City from the south and to link up with northern railway companies such as the Great Northern Railway and the Midland Railway both for passenger and coal transport. This was achieved when, on 21 December 1864, the LCDR Bridge across the Thames was opened.

The 933-foot bridge was designed by Joseph Cubitt to align with his Blackfriars Bridge. It consisted of a wrought-iron lattice-girder structure of five spans supported by massive cast-iron columns. The bridge rivalled Hawkshaw's Hungerford Bridge for ugliness and was much criticised. However, trains were soon running for the first time through London to the North via a station at Ludgate Hill. The Act had stipulated that the LCDR should provide 'workmen's trains' to run at fixed times and for fixed low fares. The reason was that the construction of the railway extension on the south bank had required the demolition of hundreds of workers' homes and the displacement of the inhabitants further away from their place of work in the City. The cost of the whole scheme, including railway lines, stations, bridges and compensation, amounted to £1,450,000, and this was a major cause of the company's eventual bankruptcy.

Following the construction of the LCDR's second Thames crossing, today's Blackfriars Railway Bridge, Joseph Cubitt's bridge and the Ludgate Hill Station it served became increasingly superfluous. Harsh criticisms of the bridge and the station it served were voiced in the *Railway Magazine* of February 1899: 'Probably no railway station since the iron roads were first invented has ever come in for such an avalanche of gorgeous and whole-hearted abuse as has fallen of late upon that of the LCDR at Ludgate Hill.' The line struggled on until services ceased altogether in 1971. The South Eastern Region of British Rail finally removed the superstructure of the bridge in 1985 using a 360-foot-tall crane which was towed across the North Sea from Rotterdam and then up the Thames to Bankside. Here, the crane was moored by the power station (today Tate Modern), where it towered 35 feet above the chimney. At night and at weekends it was towed to the LCDR Bridge, where it removed the lattice girders in sections in a period of less than two weeks, compared with the five months that would have been required using traditional methods.

Today, all that is left of the bridge are the massive LCDR insignia which were re-erected on the south side of the river and the headless cast-iron columns which still straddle the river between Blackfriars Bridge and the existing railway bridge. This

241

must be one of London's strangest and most inspiring sights, but it will soon be no more if the Thameslink 2000 planning application is allowed to proceed in its entirety. This envisages widening the existing railway bridge using the old LCDR cast-iron columns and building a new platform roof canopy spanning the river, with a station entrance on the south as well as the north side. The project is planned to take place between 2007 and 2011, but no firm decision has been taken at the time of writing.

Blackfriars Railway Bridge

In 1881, the LCDR succeeded in obtaining another Act of Parliament authorising them to build a second bridge to a new station to the east of Ludgate Hill Station. The station was called St Paul's and provided services to the Continent as well as for through traffic to the North and commuters. The bridge was designed by John Wolfe Barry, who later did the engineering design for Tower Bridge, and H.M. Brunel, the second son of

Blackfriars Railway Bridge with the headless columns of
the former London, Chatham and Dover Railway Bridge

Isambard Kingdom Brunel. It was completed in 1886 as a five-span structure with piers placed to align with the two earlier bridges to the west. The ironwork was produced by the Thames Ironworks & Shipbuilding Company of Blackwell and installed by the contractors, Messrs Lucas & Aird. The bridge carried seven lines across the river on its 81-foot-wide railroad but fanned out on the north side to a width of 123 feet to allow for the platforms of St Paul's Station, which extended over the river. Because of the great width of the bridge, it was not possible to use simple lattice girders, as with the first LCDR bridge. Consequently, wrought-iron arched ribs were built under the rails, giving sufficient support and also enhancing the bridge's appearance.

Today, Blackfriars Railway Bridge provides the only through service between north and south in central London. The old St Paul's Station was rebuilt in 1977 and renamed Blackfriars Station. Its aesthetic merits are limited to the views upstream

Old station plaques preserved in the rebuilt Blackfriars Railway Station

and downstream from the southern ends of the platforms, and the stone blocks inscribed with the names of the stations that used to be served from St Paul's Station. These inscribed stones were preserved from the old station and contain such idiosyncratic juxtapositions as Sittingbourne and Marseilles, Sheerness and Vienna.

CHAPTER 12

Millennium Bridge

The Millennium Bridge was opened to great acclaim in June 2000, but, because of the famous wobble, it had to be closed after just two days. It was reopened in February 2002 after protracted remedial work finally cured the problem of the wobble. The bridge provides a pedestrian crossing between St Paul's Cathedral on the north side of the river and Tate Modern on the south. It was the first entirely new bridge to be constructed across the Thames in London for over a century, since Tower Bridge was completed in 1894. As a footbridge, its modern, minimalist appearance could hardly be more different from the flamboyant Gothic extravagance of its nineteenth-century predecessor.

The first proposal for a bridge at this location was made in 1854, in the report of the Select Committee on Metropolitan Bridges. The Committee recommended four new bridges, including one at St Paul's, as being necessary for 'providing further means of communication across the river'. The other recommended locations were at the Tower of London, Lambeth and Charing Cross. No action was taken on the St Paul's proposal until in 1911 an Act of Parliament authorised the City Corporation to construct a new road crossing at this point as well as to rebuild the old Southwark Bridge. As described in Chapter

13, the Southwark Bridge reconstruction project, which it was estimated would cost £300,000, went ahead. The St Paul's project, however, was delayed because of the high estimated cost of well over one million pounds and because the First World War started before the exact route could be agreed. After the war, the City Corporation spent over £1,500,000 on purchasing property in preparation for building the approach roads for St Paul's Bridge but abandoned the project in 1931.

The idea of constructing a bridge at St Paul's lay dormant until the 1990s. By then, huge changes had taken place on the south bank of the river. The most significant event was the opening in 1960 of Bankside Power Station, designed by Giles Gilbert Scott, who had also designed Battersea Power Station and Waterloo Bridge. The power station, with its single tall chimney, stood across the river exactly facing St Paul's Cathedral, which for so many years has been seen as the symbol of London. Today, this seems a most extraordinary juxtaposition. By the 1970s, it was no longer considered desirable to pollute London's air by electricity generation, and by 1981 both Bankside and Battersea power stations were closed. Around this time, the wharves along the riverfront were being abandoned as the Port of London ceased to attract industrial and commercial river traffic, and the area became almost derelict.

The 1980s saw the start of a remarkable regeneration of both the disused docks to the east of Tower Bridge and of Southwark's riverside. This culminated with the decision in 1992 to transform Bankside Power Station into the Tate Modern art gallery. Now that Southwark's attractions were more on a par with those on the north bank, the idea of providing a new crossing in this area was resuscitated. Indeed, there was even a suggestion of the crossing being lined with houses, and in 1996 the Royal Academy held an exhibition of seven designs for a habitable bridge at Temple Gardens. One of the plans was by the charismatic bridge designer Santiago Calatrava, but even he could not inspire anyone to foot the bill for such an ambitious project.

In the Millennium Bridge official publication, Deyan Sudjic

describes how the much more practical and ultimately successful scheme for a footbridge was initiated at about the same time as the RA exhibition.[45] The driving personality was David Bell, managing director of the *Financial Times*, the head office of which had recently moved to the south bank by Southwark Bridge, from where there is a clear view of both St Paul's Cathedral and Tate Modern. His initiative was strongly supported by others, including Nicholas Serota, director of Tate Modern, and Southwark Council, both of whom saw the advantages of improving access to the south bank for the wealthy inhabitants on the north bank as well as City workers and tourists.

Needless to say, there was some opposition. English Heritage was concerned that a bridge would obstruct views of St Paul's from passing river traffic. Eventually, it was persuaded that, with a projected four million annual pedestrian crossings, many more people would have an amazing view of the cathedral from the bridge than would have their views impaired from the relatively few boats. The City Corporation was at first lukewarm about the project but later agreed not only to make a large contribution to the cost of construction but also to take over the maintenance. The Millennium Bridge thus became the fifth bridge to be financed as part of the Bridge House Estates, following in the footsteps of London, Blackfriars, Southwark and Tower bridges.

As has always been the case with proposals to bridge London's river, the process of obtaining authorisation and finance proved complex. The first step was to hold an architectural competition. No fewer than 227 entries were submitted for the prestigious project. A shortlist of six was selected by the distinguished panel of judges, which included Jacques Herzog of Herzog & de Meuron, the firm of architects responsible for transforming Bankside Power Station into Tate Modern. Finally, in December 1996, it was announced that the winners were the team of Norman Foster, Anthony Caro and the engineering firm Ove Arup & Partners. This was considered a dream-team combination of architecture, sculpture and engineering.

Rather surprisingly, the initial concept seems to have come not from the architect or sculptor but from Chris Wise of Ove Arup. Before the first team meeting, he was sitting in a restaurant with a colleague and did a simple drawing on a napkin of a straight line linking two banks of a river. This was to become today's slender structure, which provides minimal obstruction to views along or over the river by day, allows plenty of headroom for shipping to pass underneath and by night dissolves into the justly named 'blade of light'. Of course, a huge amount of design work was required to turn the concept into reality.

The final design is in fact a suspension bridge, although it does not conform to most people's idea of what a suspension bridge looks like. Interestingly, Norman Foster's offices overlook Albert Bridge, which does look like a suspension bridge but is in fact a cable-stayed bridge. In the case of the Millennium Bridge, there are no tall river-towers supporting curved chains, but the cables are stretched from the abutments on each side of the river and over the 'V' shaped top ends of the river-piers, hardly rising above the level of the deck they support. It is rightly described as a 'flat suspension bridge'.

The innovative design helped secure the backing of the City Corporation as well as of the Millennium Commission and the Cross River Partnership. Altogether, 18 million pounds was raised for the design and construction of the Millennium Bridge. No Act of Parliament was required because the Port of London Authority granted a licence for a river crossing and the local authorities concerned gave planning permission for the riverside developments. Contracts were awarded to Monberg & Thorsen and Sir Robert McAlpine, firms which had already collaborated on restoration work on the Forth Road Bridge, compared with which the Millennium Bridge must have seemed like a minnow. Work started in April 1999, when the project was launched by the Deputy Prime Minister, John Prescott, and proceeded with no more than the usual problems until it was almost ready for the Queen's official opening on 9 May 2000.

After attending a special service at St Paul's Cathedral, where

The Millennium Bridge under construction

she shook hands with the arch-republican Mayor of London, Ken Livingstone, the Queen walked onto the bridge but could only go halfway across because work was not fully completed. She also awarded Anthony Caro the Order of Merit, presumably in recognition of his past achievements, as it seems his only visible contribution to the final design of the Millennium Bridge is some fine but unrelated sculptures at its north end.

Despite the problems with the opening ceremony, the press and public greeted the new bridge with acclaim. According to Arup, it has the longest central span of any footbridge in the world, and everyone was eager to enthuse about a British millennium success after the despondency arising from the experience of the controversial, beleaguered Millennium Dome project. Crowds flocked in excited anticipation to the official public opening on 10 June 2000. There was a charity walk in the morning, at which a slight wobble was noticed. However, the experience of crossing the new bridge proved even more exciting than had been expected as the structure started to wobble more and more violently when it was thrown open to one and all at lunchtime. There was no panic. Indeed, many of those who were at the opening clearly enjoyed themselves and were rather disappointed when the wobble was eventually cured.

Although the engineers were sure that there was no danger of the bridge collapsing, they decided to close it while they investigated and considered what to do. The great British success story had turned into a nightmare, and the media had a field day. Matters were only made worse when the designers claimed that the wobble was caused by people walking in step, which seemed to put the blame on the public. People pointed out that Norman Foster's office was near Albert Bridge, at the approach to which there is a clear sign demanding that troops should break step when crossing. Carlton TV broadcast a documentary claiming that the designers had failed to learn the lessons of other problem bridges. Only six months before the Millennium Bridge was opened, the Pont Solférino in Paris had developed a sideways sway and was closed for repairs. An article in the *New Scientist* of 31 March 2001 cited a paper by Yozo Fujino from the University of Tokyo which described a similar effect on the Toda Park footbridge, which swayed alarmingly in 1994 when crowds swarmed onto it to watch the boat races on the river underneath.

Arup's official explanation for the wobble was that it resulted from an initial slight sideways movement caused by a chance correlation of people's steps when a considerable number of pedestrians walk along a long narrow bridge. Once this happens, it becomes more comfortable for people to walk in step and this further exaggerates the wobble. The effect is known as 'synchronous lateral excitation', and Arup even developed an equation ($F = K \times V$) to allow them to calculate the critical mass of pedestrians needed to cause this to happen. Unfortunately, it seems that similar previous bridge problems had not been widely publicised and therefore no standard tests had been developed to evaluate the stability of a footbridge when people walk in step. Arup had performed all recommended theoretical tests, and experts confirmed that according to these the bridge should have performed perfectly well. The *New Scientist*, however, was not impressed and strongly criticised the reluctance of engineers to publicise their failures or research the literature. This compares with the situation in the aviation industry where every manufacturer is informed about crashes as a matter of routine.

After much internal discussion, Arup developed a solution involving the installation under the superstructure of shock absorbers: 37 viscous dampers and 50 tuned mass dampers. The cost of the repairs amounted to five million pounds. Who came up with the money remains a secret, but it was made clear that no public finance was involved. Because of long arguments about who should pay, the need for thorough testing of the solution, and delays in the supply of the dampers, it was 18 months before the Millennium Bridge was reopened. While work progressed, passers-by may have noticed small bundles of straw hanging from under the footway. This traditional method of warning shipping that work was going on overhead arises from the time when barges bringing straw from the west for London's horses passed under a bridge being repaired and would often leave a wisp of straw hanging from the scaffolding. The old-fashioned warning sign is still used and is part of every waterman's exam, but it did seem incongruous on such a modern structure.

The bridge was finally reopened on 22 February 2002, just before the Golden Jubilee Bridge at Charing Cross. Despite being two years late, it retained the name Millennium Bridge. As reported in the *Daily Telegraph* of 23 February 2002, the reopening was performed with some hilarity. Emma Drakes from Highbury, aged 91, was given the honour of crossing first, as she had been disappointed to just miss the opportunity of crossing it in 2000. *The Sun* sent a page-three girl in a micro-bikini and carrying a plate of jelly to test out what would wobble. To most people's disappointment, there was no discernable movement, and eventually the bridge will doubtless lose its nickname of 'the Wobbly Bridge'. Sir Norman Foster, however, was delighted, as he had promised to throw himself off the bridge if it ever wobbled again.

Today, the Millennium Bridge is a popular crossing and fulfils the original concept of linking the north and south banks of the Thames at the point of two of London's most iconic buildings – St Paul's Cathedral and Tate Modern. It could hardly create less obstruction to the views of either of these buildings. For those who wish to be reminded of the wobble, it is just possible to discern dampers lurking in pairs beneath the footbridge deck. Sadly, Tony Fitzpatrick, the engineer responsible for the damper solution to the problem of the wobble, died in a cycling accident in July 2003. Thanks to his elegant design, the dampers in no way impinge on the slim, elegant lines of London's newest river bridge.

CHAPTER 13

Southwark and Cannon Street

This chapter covers Southwark Bridge and the adjacent Cannon Street Railway Bridge. Southwark Bridge was completed in 1921, replacing the original Southwark Bridge of 1819, designed by John Rennie. Cannon Street Railway Bridge was opened in 1866 to carry the trains of the South Eastern Railway to Cannon Street Station. The bridge was widened in 1893.

Southwark Bridge

The history of Southwark goes back nearly two millennia; a settlement grew up around the southern approaches to the wooden bridge which the Romans built to cross the Thames near the site of today's London Bridge. By Tudor times, Southwark had developed into the entertainment district for the inhabitants of the City, with a reputation for drunkenness and debauchery. The area abounded in brothels, inns, bear-baiting establishments, theatres and prisons, which the City liked to keep at arm's length even though this meant crossing the river to enjoy its pleasures. Until the construction of Blackfriars Bridge in 1769, the only permanent crossing in the area was provided by Old London Bridge, which had become hopelessly congested. The other method of crossing the river was to hire a wherry, with the dual risks of capsizing in the fast-flowing tidal river and being forced to pay over the odds by unscrupulous watermen. By the eighteenth century, Southwark was no longer the theatre district of Shakespeare's time but had become a working-class area the inhabitants of which needed to cross the river to their places of employment on the north side. Since the distance between London Bridge and Blackfriars Bridge is about one mile, there seemed to be a case for constructing a new bridge in between them to cater for the growing cross-river traffic requirements.

In 1811, the Southwark Bridge Act was passed, incorporating the Southwark Bridge Company and authorising it to raise £300,000 to construct a bridge from Queen Street on the north bank of the river to a site on the south bank where no through roads existed. The prospectus emphasised the benefits of such a bridge to the City, to the Borough of Southwark and to the subscribers. It described how the new bridge would relieve the congestion on London Bridge and thus facilitate commerce on both sides of the river. It would result in the construction of a handsome street from Bankside to St George's Church and therefore 'be the means of adding to the Borough a neighbourhood of inhabitants of respectability in the room of those whose removal it will occasion of an inferior class and thus increase the trade and comfort of Southwark'. The final part of the prospectus attempted to estimate the likely return to investors, based on what was claimed to be a conservative assumption that only one-sixth of the traffic over London and Blackfriars bridges would be diverted to the new, more convenient crossing. The calculations showed that subscribers could expect their shares to yield an income of 10 per cent from the tolls.

The Southwark Bridge Company engaged John Rennie as its engineer. His design could hardly have been more different from that of his elegant stone Waterloo Bridge, which was being constructed at the same time. Since the City and the Thames Conservancy had opposed the new bridge on the grounds that it would obstruct navigation, the Act stated that the arches should be as wide as possible. Rennie extended the state of the art of bridge building by designing three flat cast-iron arches supporting a 42-foot-wide roadway. The centre arch, at 240 feet, was the longest of its kind ever constructed, and doubts were expressed about the stability of such a long arch. The design was examined by the eminent scientist Dr Thomas Young, who concluded that the basic design was safe but that every detail of materials and workmanship should be done with skill and accuracy.

In his autobiography, Rennie's son, also called John, described the difficulties he encountered as his father's chief assistant.[46] His most dramatic account was of the efforts involved in acquiring and transporting the massive 20-ton granite blocks needed to encase the river-piers. He travelled to Aberdeen only to be told that such a project had never been done in Britain before and no one was willing to try. Nothing daunted, Rennie went 30 miles north to a quarry where the local workmen agreed to hew a massive 25-ton block of granite, with the encouragement of ample wages and a good supply of whisky. They loaded the block onto a cart which was eventually set in motion by the use of 14 horses. A further problem arose when they reached the turnpike and the toll-keeper did not know what toll to charge since he had never seen such a mass of stone before. Eventually, a negotiated settlement was reached and the block arrived at the nearby port of Peterhead. Here, Rennie had to agree to indemnify the ship's captain for any loss incurred on the journey to London.

The supply of the ironwork proved an equally challenging task. Walker Bros of Rotherham was the only firm considered capable of meeting the stringent requirements of Rennie's design, having made many of the cannons used in the Peninsular War against Napoleon. Thousands flocked to Rotherham to watch the construction of the gigantic cast-iron structures which the locals called a wonder of the world. The ironwork was successfully delivered, but the effort bankrupted the firm.

Construction work on Southwark Bridge progressed slowly, partly due to the war against France, which lasted until the eventual defeat of Napoleon at Waterloo in 1815. As usual, a number of accidents occurred, including one disastrous incident resulting in the loss of 13 lives. In a letter to *The Courier* of 5 October 1816, John Rennie Jr described how 15 workmen hailed a boat when they had finished working on the bridge. Two watermen approached and the whole party jumped into their boat, despite attempts by the watermen to stop them. The strong tide caused the boat to crash into a barge, and when the boat upset, everyone was precipitated into the water. A passing police boat saved two of the workmen, and the watermen, who were strong swimmers, managed to reach the riverbank. The rest were not found at the time and were presumed drowned. Four of the bodies were later recovered. Rennie added that 'the men have been regularly warned about rushing into the boats, but to no purpose'.

The bridge was eventually opened without ceremony on 24 March 1819, as the clock of nearby St Paul's Cathedral struck midnight. Since the prospectus had grossly underestimated the cost, which finally amounted to a staggering £800,000, there was no money left to pay for an opening spectacle. Already in 1813 one of the subscribers had warned that the bridge was likely to be a financial disaster.[47] He claimed that the prospectus was misleading because the large number of people living on the south side of the river in Surrey and Kent would prefer to continue using Blackfriars or London bridges, which were by now free from toll. Therefore the estimate that Southwark Bridge would attract one-sixth of the traffic, despite the need to pay tolls, had been pure guesswork. On this basis, the subscriber demanded the right to pay a forfeit in order to withdraw from the undertaking. Needless to say, this request was not granted.

In his anger at the prospect of losing money on his shares, he objected to Rennie's design as

> the introduction of a kind of filigree edifice between two
> such noble structures as London and Blackfriars Bridges.
> The want of uniformity in our squares and streets has
> long been the subject of censure on the part of foreign

architects. The new iron interloper spoils the noble succession of bridges including the new Strand Bridge.

Interestingly, the Strand (soon renamed Waterloo) Bridge was also designed by Rennie, and in a few years' time Rennie's granite structure would also replace the 600-year-old London Bridge.

The forecast of a financial disaster proved correct, because very few people used the bridge. The holders of £150,000 preference shares received an average of 1.5 per cent per annum, while the holders of the ordinary shares never received any dividends. The City Corporation conducted lengthy negotiations with the proprietors about purchasing the bridge and freeing it from toll. Initially, the sum of £230,000 was offered, but the proprietors wanted more. However, the financial situation continued to worsen and eventually, in 1868, the proprietors were forced to accept an offer of £200,000, which represented a considerable loss on the original cost of £800,000.

Even after Southwark Bridge was freed from toll, its usage was limited, partly because of the steep gradient of 1 in 18, which

John Rennie's Southwark Bridge of 1819

resulted in a hump, causing difficulties for horse-drawn traffic, and partly because the northern approach road was not convenient. The *Daily Chronicle* of 8 August 1903 reported that 'you may live a long time in London without crossing Southwark Bridge and frequently it has an air of solitude'. Southwark Bridge seems also to have been almost ignored by artists and writers. The *Daily News* of 2 June 1921 stated that it had the fewest historical and literary associations of all London's bridges. There are, however, some references to be found in Dickens' works. He called it 'the iron bridge' in *Little Dorrit*, and it features in the opening scene of *Our Mutual Friend*, in which Gaffer Hexham searches for dead bodies in the river from his boat as he 'floated between Southwark Bridge which is of iron and London Bridge which is of stone, as an autumn evening was closing in'. At that time, Cannon Street Railway Bridge did not intervene.

The City debated long and hard about what to do with the languishing Southwark Bridge, and several reports were issued by the Bridge House Estates Committee. Many considered that the best solution would be to build a new crossing at St Paul's Cathedral, but the estimated cost of new approach roads at that location amounted to over one million pounds. In 1911, the City submitted a Bill to Parliament asking for authority to rebuild Southwark Bridge for £261,000, as well as to construct a totally new St Paul's bridge. In view of criticisms of the aesthetic merits of certain late-nineteenth-century crossings, it was recommended that the best architectural advice be sought in both cases. As we have seen, the St Paul's bridge remained on the drawing board. For Southwark Bridge, the City engaged the architect Sir Ernest George (1839–1922) to design a 55-foot-wide bridge with no hump.

His steel bridge consisted of five arches instead of the three of Rennie's Southwark Bridge. The main reason for this was that the five-arch Cannon Street Railway Bridge had been built 150 yards downstream of Southwark Bridge. The long chains of barges that plied the Thames had to twist their course with dangerous alacrity because the piers of the two original bridges

Southwark Bridge today

were not aligned. George designed stone river-piers with pierced lunettes, carried up above the steel arches, giving character to the architecture and providing recesses on the footways. He worked with the engineering firm Mott, Hay and Anderson on the design. The £278,148 construction contract was awarded to Sir William Arrol & Co., who had also supplied the steelwork for Tower Bridge.

Work started in 1913 but was interrupted by the First World War. The Government commandeered the steel for military purposes but allowed work on the piers to continue so that a temporary superstructure could be added in the event of damage to other bridges. As it happened, although bombs did fall near by, no fatal damage was done to any of London's Thames bridges in either of the world wars.

Southwark Bridge was finally completed in 1921, at a cost of £375,000. Since all the finance was provided by the Bridge House Estates, the public was not concerned at the cost overrun of nearly £100,000. The bridge was opened by George V and Queen Mary on 6 June. In view of the less than glorious history of Rennie's Southwark Bridge, many suggestions were made to change the name. The favoured options were the King George or the Victory Bridge. Although the Strand Bridge had successfully been renamed Waterloo after the victory over Napoleon, the record of naming bridges after kings or distinguished people was not good. Examples include the Edward VII Bridge, which was soon known as Kew Bridge; the Victoria Bridge, which was later renamed Chelsea Bridge; and the William Pitt Bridge, which was changed to Blackfriars Bridge when Pitt lost his popularity. George V himself sensibly decided to keep the old name. This decision was welcomed by *The Times* of 7 June 1921, which pointed out that the new bridge had been constructed during time of war just like its predecessor and that it was better to maintain tradition with such an important structure as a Thames bridge rather than indulge the fancy for some 'new-fangled nomenclature'.

The opening ceremony was a festive occasion, with crowds massing on either side of the river. The Lord Mayor welcomed the royal party with a loyal speech, but the Mayor of Southwark was not to be outdone. When George V crossed to Southwark, the Mayor reminded him that 'James I of Scotland was married at the Cathedral Church to Joan, niece of Cardinal Beaufort. At our local church of St George the Martyr, Henry V sought the Divine Blessing when setting out for his victorious campaigns in France.' Having established Southwark's royal connections, he handed over to the Bishop of Southwark, who pronounced a blessing on the bridge and those who had designed and built it.

The new bridge provides a carriageway of 35 feet and two 10-foot-wide footways, which are cantilevered out from the supporting arches. The cantilevering was required because Rennie's original 42-foot-wide abutments were used for the 55-foot-wide bridge. The river-piers had to be rebuilt in different places because they now had to support five arches instead of three. With the wider carriageway and the removal of the hump, it was expected that usage of the crossing would increase substantially. In fact, according to Transport for London statistics, Southwark Bridge is still by far the least used of all

London's Thames bridges, with fewer than 13,000 crossings per day. Because of the new attractions on the south bank, probably more people today walk along the Thames Path and pass under the arches of Southwark Bridge than cross it. Those who go under the southern approach arches will see slate murals showing pictures of the old Frost Fairs. Under the arches on the north side are more traditional tile murals showing pictures of the old Southwark Bridge and its construction.

On the night of 20 August 1989, Southwark Bridge was the scene of one of London's most disastrous river collisions, when the *Bowbelle* dredger crashed into the stern of the *Marchioness* pleasure cruiser, resulting in the loss of 51 lives. The *Marchioness* had set out at about one in the morning from Charing Cross with 113 young people on board, who were celebrating the birthday of Antonio Vasconcellos, a Portuguese financier. After passing through Blackfriars Bridge, the *Marchioness* gradually overtook another pleasure cruiser, the *Hurlingham*, and then sailed under the centre arch of Southwark Bridge. Although one of the reasons for rebuilding Southwark Bridge with five arches was to ensure alignment with Cannon Street Railway Bridge, it seems that the centre arches are still slightly offset and the *Marchioness* had to adjust its course between the bridges. Meanwhile, the *Bowbelle* had passed the *Hurlingham* but seemingly was unaware of the presence of the *Marchioness* as it too passed through the centre arch of Southwark Bridge and smashed into the stern of the much smaller vessel. Even though both boats were travelling in the same direction at no more than 8 knots, the massive weight of the 260-foot-long dredger caused the *Marchioness* to tip over and sink. The *Bowbelle* itself bumped into Cannon Street Railway Bridge after the collision but none of its nine-man crew was injured.

Fifty-one people on board the *Marchioness* drowned, including the skipper, Stephen Faldo. This death toll was in spite of the efforts of the crew of the *Hurlingham*, which arrived soon after the accident and was able to help in the rescue efforts. The coroner issued a verdict of 'unlawful killing'. Allegations were made about heavy drinking among the crew of the *Bowbelle*.

However, the Crown Prosecution Service decided that there was not sufficient evidence to bring any prosecutions, although many found it hard to believe that a Thames pleasure cruiser could have been invisible to the approaching dredger. Later inquiries severely criticised the state of river safety. Appropriately, Southwark Cathedral has a moving memorial to the victims of the tragedy.

Cannon Street Railway Bridge

Cannon Street Railway Bridge crosses the river to Southwark only 150 yards downstream of Southwark Bridge. Up to medieval times, the River Walbrook used to flow into the Thames here from the north. John Stow states that the Walbrook had been largely covered over by the end of the sixteenth century,[48] and today, like the other 'hidden rivers' of London, it has been turned into an overflow sewer. This was also the location of the Steelyard, where merchants of the Hanseatic League had special privileges to run their own affairs after Richard I granted them a charter allowing them to trade throughout the country in 1194. The merchants brought in much trade, to the delight of later monarchs who benefited from the customs duties. Their success allowed them to build an impressive mansion called Steelyard Hall, for which Holbein painted two huge pictures, *The Triumph of Riches* and *The Triumph of Poverty*, both unfortunately now lost. However, the merchants aroused the jealousy of the English trading guilds and were eventually banished by Elizabeth I. The site of the Steelyard was built over in 1865 to form Cannon Street Station. According to Stow, the name Cannon Street has nothing to do with armaments, but is derived from the former Candelwykestrete, where candle-makers dwelt. Today, the tradition of candle-making is kept up by the Tallow Chandler's Hall, which occupies a site in Dowgate Hill next to the station.

After the London, Chatham and Dover Railway had started building their terminus on the north bank at Blackfriars, the South East Railway were determined to match them by bringing their trains even nearer to the centre of the City. In June 1861, they obtained permission to extend their railway line across the Thames and to construct a station on the north bank within walking distance of the Bank of England. The SER engineer Sir John Hawkshaw designed both the bridge and the station. The station was a typically splendid triumph of Victorian engineering, stretching from Cannon Street to the banks of the Thames. The platforms were covered by a vast iron roof, which had a span of 190 feet. It was glazed over two-thirds of its surface. The roof was supported by solid brick walls and enclosed by two towers of monumental proportions, which seem to mimic church steeples by Sir Christopher Wren. The towers contained water for use in running the hydraulically powered lifts and for cleaning the trains.

Compared with the magnificent station, Cannon Street Railway Bridge was strictly functional. It originally took nine lines across the 855-foot width of the Thames. Its five spans of wrought-iron plate girders were supported by piers of four cast-iron columns. On either side of the tracks, there were narrow footways. One was used only by railway staff, but the other was open for pedestrians for a halfpenny toll until 1877, when the Metropolis Toll Bridges Act abolished all tolls on London's

1864 view of Cannon Street Railway Bridge
with the original massive station arch

bridges. Unlike at Charing Cross, where pedestrians could use the footway after the Metropolitan Board of Works paid compensation to its owners for the loss of the right to charge tolls, the SER decided to close the Cannon Street footway.

The station and bridge were opened on 1 September 1866. The total cost was £350,000, which stretched the SER finances to such an extent that they were unable to build the essential railway hotel. This project was soon taken on by an independent company, who commissioned E.M. Barry to design a grand building in a similar style to what he had already constructed for the SER at Charing Cross Station.

Most of the railway traffic originally ran from London Bridge across the river to Cannon Street and then back over the river to Charing Cross. This involved complicated manoeuvring, as the trains had to back out of Cannon Street Station for the southern crossing. When the District Underground line was constructed shortly afterwards, providing a direct service from Cannon Street to Embankment, the overland trains to Charing Cross ceased to attract many customers. However, it seems that prostitutes found the first-class carriages, with at least seven minutes of uninterrupted journey time from Cannon Street to Charing Cross, highly profitable.[49] The fare was considerably less than the charge for services rendered. Evidently, drawn blinds on this line were a common sight until the service was withdrawn in 1916.

Mainline traffic in general, however, did increase, and in 1893 the bridge was widened to take ten lines. This involved building two extra sets of columns on the upstream side, while leaving the original four downstream sets of columns in place. At the time, it was claimed that Cannon Street was the widest railway bridge in the world. Its place in history was confirmed when various revolutionary bodies met there in 1920 to form the Communist Party of Great Britain.

During the Second World War, Cannon Street Station suffered massive bomb damage on the night of 10 May 1940. The hotel was completely gutted and the station roof nearly collapsed. The glazing was removed for safety, but the roof proved too weak to reinstall the glass after the war. This is

264

probably the origin of an amusing but apocryphal story often told on pleasure-boat trips as they pass under the bridge. According to this story, for security reasons the station roof was removed during the war to a location in the country. Unfortunately the location was bombed and the roof destroyed, while Cannon Street Station itself remained intact. In fact, the damaged roof remained *in situ* without being reglazed, until it was demolished in 1959 and sold for scrap. The Royal Fine Art Commission insisted that the towers and brick walls were preserved, and they still stand as a reminder of the once magnificent train shed. In 1991, a modern air-rights office was built, which juts out rather incongruously between the two Victorian towers. Here the traders of the London International Financial Futures Exchange operate and can enjoy views of the river from the roof garden on top of the building.

Cannon Street Railway Bridge itself lost its only decorative features during strengthening work in 1981, when the Doric capitals of the columns were encased in concrete. Today, it has surely replaced Hungerford Bridge as London's ugliest Thames crossing.

Cannon Street Railway Bridge today, with the original station towers

265

CHAPTER 14

London Bridge

London Bridge crosses the Thames from Southwark on the south, one of London's poorest districts, to the very heart of the City's financial district on the north. It is the latest of several crossings that have spanned the river in this area. Because of its strategic position, there has been a river crossing here since Roman times, and indeed London Bridge was the only crossing in central London right up until 1750 when a further bridge was built over the Thames at Westminster. Southwards, the bridge leads to Borough High Street, the ancient route to the south coast, and passes Southwark Cathedral on its upstream side. Northwards, it leads to King William Street and then on to the Bank of England, after passing by the Church of St Magnus the Martyr on the downstream side. It seems appropriate that today's London Bridge is still associated with these two outstanding ecclesiastical buildings, since in its most famous incarnation as an inhabited bridge it was built by a priest.

Of all the Thames bridges, London Bridge has played the most outstanding role in the life of London, and also of England, for nearly 2,000 years. Although there is no archaeological evidence, it is likely that the Romans built a wooden bridge in this area soon after they conquered Britain in AD 43. At that time, the Romans chose Camulodunum, today's

Colchester, as the capital of Roman Britain, as it was a major tribal centre and allowed them to control the east of England, where the most powerful tribes were based. In order to secure the route over the Thames to Camulodunum from the south, the Romans founded Londinium as a settlement on the north side of the river, with a smaller settlement on the south side. Soon, a road network converged at this point, leading from Richborough and Chichester in the south to Camulodunum. Later, the great roads of Watling Street and Ermine Street (to Chester and York respectively) were constructed from Londinium to improve the legions' access to the North.

This bridge would not have survived the revolt of Boudicca, Queen of the Iceni, who in AD 61 burned down Camulodunum and Londinium, and slaughtered all the Romans she could find while the main Roman legions were pursuing their conquest of the north of Britain. Boudicca was exacting her revenge on the Romans, who flogged her and raped her daughters after she protested at their seizure of her property following the death of her husband. Following the revolt, the Romans marched their legions back to Londinium and, with relentless efficiency, defeated Boudicca and her marauding tribesmen. Londinium was rebuilt and soon superseded Camulodunum as the Roman capital of Britain.

The first Thames crossing for which archaeological evidence exists was built downstream of the present site of London Bridge, probably in AD 80–90. It was a wooden bridge with a central drawbridge to allow taller ships to pass through. A model of this, based on research by MoLAS, is displayed in the Museum of London.

Nothing is known of what happened to London Bridge after the Romans withdrew from Britain in 410. The Saxons who conquered Britain after the Romans had left made their main settlement further west, in the Strand area, leaving Roman Londinium to decay. Since a wooden bridge requires considerable maintenance, it is likely that it fell into disrepair and that no permanent crossing existed for several centuries, until Alfred the Great resettled the Roman city in the ninth century. References to a 'London Bridge' start to appear in tenth-century documents, at which time the Vikings from Denmark were a constant threat. The most famous event concerning London Bridge before the Norman conquest of 1066 occurred in 1014. As related in the Olaf Sagas,[50] the Danes had occupied London and the Saxon King Aethelred, allied with Olaf, King of Norway, was trying to recapture his capital city. As the Danish forces stood on London Bridge to face the attack from the south, King Olaf sailed his fleet up to the bridge, tied ropes around the supporting wooden piles and rowed as fast as he could down the stream. The bridge collapsed and a great part of the Danish army fell into the river and drowned. A poem was composed by the Norse poet Ottar Svarte to commemorate the battle. The poem starts with the lines:

> London Bridge is broken down,
> Gold is won and great renown.

The well-known children's nursery rhyme could have its origin in this poem, although the modern version does not appear until the seventeenth century:

> London Bridge is falling down,
> Falling down, falling down.
> London Bridge is falling down,
> My fair lady.

The introduction of the 'fair lady' is obscure, although it has been suggested that this refers to Eleanor of Provence, who was given control of the later, stone London Bridge by Henry III in 1269. She collected the rents but failed to maintain the bridge properly, and so it fell into disrepair.

By the middle of the twelfth century, the wooden London Bridge had been repaired or even rebuilt on several occasions, partly because of the force of the tidal river and partly because of frequent fires. The final rebuilding occurred in 1163 under the control of Peter de Colechurch, who was the priest at St

268

Mary Colechurch in Cheapside, medieval London's main shopping street. Having experienced the problems of maintaining the wooden bridge, Peter de Colechurch decided that the time had come to construct a stone bridge befitting London's growing importance as capital of England and as an international trading centre. The first pile was laid in 1176, and the priest was to devote the rest of his life to the groundbreaking project of spanning the fierce tidal flow of the nearly 1,000-foot-wide river.

It may seem surprising today that a priest should instigate and manage the construction of a bridge, but in the Middle Ages bridge building was seen as an act of piety. The famous Pont d'Avignon across the Rhône was built by St Bénezet at about the same time as the stone London Bridge (known today as Old London Bridge), and both bridges had chapels on them. Old London Bridge's chapel was dedicated to the Archbishop of Canterbury, St Thomas Becket, who had fallen out with Henry II by insisting on the rights of the Church in opposition to the monarch. Becket had been martyred in 1170 when four knights took literal action on Henry II's possibly rhetorical question 'Who will rid me of this turbulent priest?' and murdered him in Canterbury Cathedral. Sadly, Peter de Colechurch did not live to see the completion of his life's work. On his death in 1205 he was laid to rest in St Thomas's Chapel. His bridge was eventually opened in 1209 and was to last over 600 years. When it was finally demolished in 1832, some bones purported to be the remains of Peter de Colechurch were discovered in the undercroft of the chapel and were deposited in a casket in the British Museum. Unfortunately, on subsequent examination only one bone was found to be human, the rest being of animal origin.

Finance for the 33-year-long project to build Old London Bridge was raised largely from a tax on wool, which was England's most important export and formed the basis of the country's wealth. This gave rise to the saying that 'London Bridge was built upon woolpacks'. Money was also accumulated from a variety of other sources, including gifts 'to God and the

Bridge'. As time went by, an extensive property portfolio was built up by the Bridge House Estates, which was responsible for the maintenance of Old London Bridge, and in 1282 the BHE was granted a Royal Charter. The name arose from the Bridge House that became the administrative headquarters of Old London Bridge. This was situated on the south of the river downstream of Old London Bridge, next to St Olave's Church – the church dedicated to the (now sainted) Olaf who had pulled down London Bridge in the battle against the Danes in 1014. Bridge House itself has long disappeared and St Olave's Church was replaced in the 1930s by the art deco office block known as St Olaf House. Over the centuries, the BHE portfolio has grown to be worth over £500,000,000. Today, it is responsible for maintenance of all the bridges which cross the Thames from Blackfriars to Tower Bridge.

Old London Bridge was always an inhabited bridge. It consisted of 20 arches, one of which was a drawbridge, and spanned 905 feet of treacherous tidal water. There has been much speculation as to how it was constructed with the limited technology available to the twelfth-century bridge builders. Writing in the sixteenth century, John Stow's theory was that the course of the Thames had been diverted to the south while the bridge was being constructed,[51] but this idea has been rejected by later historians as virtually impossible. Excavations on the foundations carried out when Old London Bridge was demolished in 1832 showed that it was likely that the bridge was built in the river without the use of cofferdams or caissons. Progress was slow, as it involved driving short piles into the river-bed at low tide and then filling in the area around where the piers were to be built with rubble. A large piling machine was then placed on the rubble to drive in additional long piles to surround the inner piles. Finally, the gap was filled in with more rubble. The result was a pointed oblong structure known as a starling, on top of which the piers and arches could be constructed. No wonder the project lasted 33 years. Many lives would have been lost in the course of construction. Gordon Home estimated that as many as 200 men would have been killed working on the building and maintenance of Old London Bridge between the start of construction in 1176 and final demolition in 1832.[52]

The essence of Old London Bridge was its irregularity. Unlike later bridges, its arches were of many different widths and shapes. The only thing they had in common was that they were all Gothic pointed arches, apart from the drawbridge. Since the average gap between the arches was less than 30 feet, the resulting concentration of the tidal flow caused excessive strain on the starlings, which were in constant need of repair. Over the years, they tended to grow in size as more rubble was added to bolster them, and this only made matters worse. As well as causing damage to the structure, the concentrated flow of the water proved a danger to passing river traffic. The 19 starlings of the bridge acted almost like a weir, producing a drop in the water level of up to 6 feet, and it was really only safe to pass under the bridge when the flow of the water was slower, at either side of the turning of the tide. At this point, the watermen would race forward to be the first to 'shoot the bridge'.

Apart from running the risk of being capsized by the rushing water, boatmen were subject to other hazards. The starlings were submerged in water during high tide and boats could easily be washed onto their surfaces, where they could be stuck until the next tide arrived to refloat them. Less dangerous but nevertheless unpleasant was the risk of streams of urine or lumps of faeces dropping onto you as you passed under the many public and private latrines that disgorged their contents into the river from the houses above.

Many people were drowned or seriously injured when passing through Old London Bridge and this gave rise to the proverb 'London Bridge is for wise men to go over and fools to go under'. Some people sensibly refused to go under the bridge. Cardinal Wolsey, on his frequent river journeys from his palace at Hampton Court to visit Henry VIII at the royal palace of Placentia in Greenwich, would disembark from his ceremonial barge before reaching Old London Bridge. There he would mount a mule which he rode along Thames Street to

Billingsgate, where he returned to his barge to continue on to Greenwich. There is no record that Henry VIII himself avoided passing under the bridge on his equally frequent journeys from Westminster to Greenwich, but Henry was a much bolder man than the cautious cardinal.

The dangers of passing through the arches of Old London Bridge were increased in 1581 when the Dutchman Peter Morice installed a waterwheel in front of the northernmost arch in order to pump water to a cistern on the north bank and supply water to subscribers in the City. According to John Stow, he obtained permission to do this by holding a demonstration of how his wheel could pump a jet of water over the steeple of the Church of St Magnus the Martyr. His wheel was made of wooden struts and used an ingenious system of pulleys so that it automatically raised and lowered itself with the tide. So successful was the enterprise that it was incorporated as the London Bridge Waterworks. Four more wheels were later installed by the northern arches, and two at the Southwark end of the bridge. The result was to concentrate the flow of the tide at the central arches, which became even more difficult to navigate. The other effect was to increase the action of Old London Bridge as a sort of weir, allowing the river to freeze in many of the much colder winters prevalent at the time and inspiring the watermen to set up the first Frost Fairs.

A particular feature of Old London Bridge, and replicating its earlier wooden predecessor, was the drawbridge, which was sited near the middle of the bridge. The drawbridge was originally designed both to allow tall ships to pass through to land their cargoes at the upstream harbour of Queenhithe and to prevent an enemy from crossing the river to the City. An ornately decorated wooden gate, known as Drawbridge Gate, stood at its northern end. This provided a third line of defence against attack from the south, after the formidable stone gate at the Southwark end and the drawbridge itself.

Drawbridge Gate also became the site for a gruesome tradition started in 1305 when the head of the Scottish patriot William Wallace was displayed there on a spike. Wallace had

The 1814 Frost Fair with Blackfriars Bridge in the background

fought for Scottish independence against the English King Edward I, known as the 'Hammer of the Scots'. After his eventual defeat, Wallace was captured, having been betrayed by another Scottish lord. Edward I had him convicted of treason for his rebellion and condemned him to death. The punishment for treason at that time was the utterly cruel method by which the condemned man was hanged, drawn and quartered. After a torturous process, Wallace was decapitated and his body divided into four quarters. To deter other rebels, his body parts were displayed in various northern cities: his right arm in Newcastle, his left arm in Berwick, his right leg in Perth and his left leg in Aberdeen. His head was parboiled to inhibit the process of rotting and impaled on a spike on the Drawbridge Gate of Old London Bridge.

From this time until the seventeenth century, heads of convicted traitors were displayed in several places around London, including on Temple Bar, but the most distinguished or notorious found themselves impaled at Old London Bridge. Executed royals never suffered this indignity, but even the head of Sir Thomas More, one-time friend and Chancellor of Henry VIII, was put on a spike on Drawbridge Gate after he was executed in 1535 for refusing to acknowledge the King as supreme head of the Church of England. Legend has it that his daughter, Margaret Roper, prayed that his head should fall into her lap as she passed under the drawbridge and so it did. More probably, she obtained the head through bribery. She had it buried in the Roper vault in St Dunstan's Church in Canterbury.

After the demolition of the Drawbridge Gate and its replacement by Nonsuch House in 1577, heads were impaled at the Great Stone Gate at the Southwark end of the bridge. Since treason and executions were all too common, there would often be a number of heads on display. One German tourist even counted 34 heads on the Great Stone Gate on his visit to London in 1592.[53] The end of this barbaric custom came in 1678. The last head to be spiked on the Great Stone Gate was that of William Stayley, who was falsely accused by Titus Oates, the concocter of the 'Popish Plot' conspiracy, of plotting to murder Charles II with the aim of putting the King's Catholic brother James on the throne. After this, the heads of traitors were demoted to Temple Bar, where Samuel Johnson frequently noted this disgusting practice, which was finally ended altogether in 1746.

Just like its wooden predecessor which was at the centre of the battle for the control of London in 1014, Old London Bridge played its part in many historical events which decided the fate of the kingdom. One of the most surprising incidents occurred in 1216, only seven years after Old London Bridge was opened. King John, who has been described as England's worst king, was extremely unpopular because of his arbitrary rule, the continual defeats in wars against France and the high taxes imposed to finance these wars. In 1215, he was forced by the barons to sign Magna Carta at Runnymede. This famous document defined the rights and responsibilities of the monarch and of his subjects for the first time in history. However, the Pope was not pleased to hear of this challenge to royal supremacy and he declared

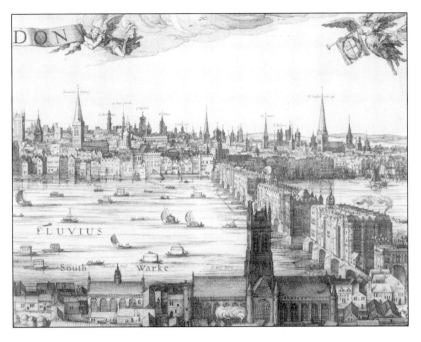

C.J. Visscher's seventeenth-century engraving of Old London Bridge

Magna Carta invalid. The barons then turned for help to Louis, the Dauphin of France, who had been at war with King John for most of his reign. They went so far as to invite Louis to become King of England in place of John. Louis was only too pleased to take over France's long-standing enemy so easily, and in 1216 he crossed the Channel with his army and marched unopposed to London. There, he was greeted with applause by the citizens as he crossed Old London Bridge on his way to St Paul's Cathedral. John died of dysentery soon afterwards and the barons had second thoughts about being ruled by a French king. They bribed Louis to renounce the throne and return to France, where he later succeeded his father as Louis VIII. John's son Henry, who was a mere child and therefore likely to be more easily manipulated than his father, was crowned Henry III. Never again would France take control of England throughout the long history of warfare between the two countries.

When he reached maturity, Henry III proved a weak king.

Like his father, he fell out with the barons, and for similar reasons. This time the barons declared war, and in 1264 their leader, Simon de Montfort, marched on London. He was quartered with his army in Southwark, but the royalist party raised the drawbridge on Old London Bridge, closed the gates and threw the keys into the Thames. However, most of the citizens of London sided with de Montfort. They managed to force open the gates and raise the drawbridge to let de Montfort's army cross the river and take control of London. In the following year, de Montfort was defeated and killed at the Battle of Evesham by the royalist army led by Henry III's son, the future Edward I. The rebellion soon ended and Henry III regained his throne.

The next major uprising resulted not from the discontent of the barons, but from the anger of the common people, who were forced to pay a series of poll taxes to finance the continuing wars with France, known as the Hundred Years War. In 1381, Wat Tyler instigated the rebellion which became known as the Peasants' Revolt. He was assisted by John Ball, a vagrant preacher who made famous the two-line verse expressing his basic principles of humanitarian egalitarianism:

> When Adam delved and Eve span,
> Who was then a gentleman?

Under the leadership of Tyler and Ball, a vast crowd of peasants and other low-paid workers from Kent marched on Southwark. There they found that the Lord Mayor, William Walworth, had raised the drawbridge to prevent them attacking the City. However, as so often, the citizens of London mainly sympathised with the rebels. When Wat Tyler threatened to burn down the houses at the Southwark end of Old London Bridge, the citizens agreed to lower the drawbridge and let the by now very threatening mob cross the river. The Kentish peasants were then joined by another crowd from Essex. No one in authority was safe. Simon Sudbury – the Archbishop of Canterbury and Lord Chancellor – had taken refuge in the Tower of London, but the constable of the Tower was a rebel sympathiser and opened the gates. The peasants seized Sudbury as he was at prayer in St John's Chapel, dragged him to Tower Hill and executed him to loud acclaim. His severed head was paraded through the streets and displayed on a spike at Drawbridge Gate.

The revolt ended when William Walworth stabbed Wat Tyler at Smithfield while the child king, Richard II, was offering to meet the peasants' grievances in a negotiated settlement. Richard then bravely offered himself as captain to the remaining rebels, agreed to abolish the hated poll tax and led them to the open countryside north of Smithfield, from where they eventually dispersed. As they crossed Old London Bridge on their way home, the men from Kent would have been devastated to see their leader's head impaled at Drawbridge Gate in place of the head of the Archbishop, which they themselves had so recently put up there on a spike.

A number of other rebellions took place during the life of Old London Bridge. Since this was for so long the only river crossing in central London, any attack from the south had to be directed towards this point. The most vicious rebellion after the Peasants' Revolt occurred less than 70 years later when the men of Kent again rose up, this time against the ineffective rule of Henry VI and the oppressive taxes raised to indulge the extravagance of his arrogant French wife, Margaret of Anjou. In July 1450, Jack Cade led a motley crowd of around 30,000 men to Southwark, where they encamped while Cade negotiated with sympathisers in the City to allow him to cross Old London Bridge. Eventually, the keys of the bridge gates were handed over and the initially peaceful rebels entered the City. Cade submitted his demands to the King in writing, so he must have had a number of more educated supporters.

Once in control of the City, however, the mob became increasingly violent. They captured and beheaded Lord Saye, who had become the focus of their hatred as a leading landowner in Kent. Following increasing incidents of looting and rape, the citizens of London turned against the invaders. When the mob retired to Southwark for the night, the

Londoners gained control of the bridge. When Cade found out that the bridge was barred against him, he attacked with full force and a fierce battle raged all night. Many of the bridge inhabitants perished as their houses were burned down while they were trapped inside. By morning, the mob had been repulsed, and the rebellion ended with Jack Cade's death in a skirmish after he had been pursued into Kent on the following day. His head was severed from his corpse and, just like Wat Tyler's before, it was impaled on Drawbridge Gate.

There were two later unsuccessful attacks on Old London Bridge: one in 1471 by Thomas Fauconberg, who tried to restore Henry VI to the throne after he had been deposed by Edward IV in the Wars of the Roses; and the other in 1554 by Sir Thomas Wyatt, who opposed Queen Mary's intended marriage to the Roman Catholic Philip II of Spain. The last time Old London Bridge saw military action was during the Civil War. In 1642, Charles I set up his headquarters in Oxford while General Fairfax led the Commonwealth New Model Army to Southwark with the intention of gaining control of London. No force was needed, as Londoners supported Cromwell and the army was allowed to cross Old London Bridge peacefully. The Civil War ended with the defeat of the King, who was executed in Whitehall in 1649. No king's head has ever been displayed at Old London Bridge. Following his execution, the head of Charles I was sewn back onto his corpse before he was eventually buried in St George's Chapel at Windsor Castle.

Old London Bridge reached the pinnacle of its fame in the sixteenth and seventeenth centuries. Tourists came from all over Europe to admire what was considered a wonder of the world. It must be admitted that many visitors were fascinated by the display of heads as much as by the bridge's other charms. However, the centuries-old bridge was steeped in history, was the longest inhabited bridge in the world and possessed an extraordinary mixture of architectural styles that made it typically English at a time when the rest of Europe was embracing the classical baroque.

The most astonishing feature of Old London Bridge was the

number and variety of the houses that lined both sides of the roadway. There were other inhabited bridges, such as the Ponte Vecchio in Florence and the much later Pulteney Bridge in Bath, both of which still exist and give some idea of the appearance and atmosphere of Old London Bridge. However, none of these bridges approached the size and idiosyncratic magnificence of Old London Bridge. The houses lining the bridge were three storeys high. The ground floors housed various businesses, mainly shops, while the upper two floors and attics provided living accommodation. All the buildings apart from St Thomas's Chapel were constructed of wood in order to lighten the load on the superstructure. As the shoppers, traders and passing vehicles thronged the bridge, the overall impression was of a busy, claustrophobic and top-heavy thoroughfare. This impression was heightened because the three-storey-high buildings were supported on a narrow roadway only twelve feet wide and were built without gaps, apart from the drawbridge. The feeling of enclosure created by the tall houses lining the dark, narrow roadway was exacerbated by the fact that many houses had hautpas, wooden platforms joining the top floors of two houses on opposite sides of the road. These provided extra accommodation and also helped strengthen the buildings, which had a tendency to lean outwards over the river.

The most impressive structure on the bridge was St Thomas's Chapel, which stood on the downstream side of the bridge, resting on the wide starling of the ninth pier from the City end. From there, it rose as high as the neighbouring houses. It could be approached by boat as well as from the roadway. Built of stone in the early English Gothic style by Peter de Colechurch in the twelfth century, it was reconstructed in the perpendicular Gothic style in the fourteenth century. As a chapel, it did not survive Henry VIII's Dissolution of the Monasteries in the 1530s, especially as it was dedicated to St Thomas Becket, who had opposed an earlier King of England. The building, however, remained in a simplified form as a shop until the final removal of all the houses in 1762.

The pre-eminence of St Thomas's Chapel was succeeded by

the exotic Nonsuch House, which was constructed next to the drawbridge in 1577. Nonsuch House was named after Nonsuch Palace, the name of which implies that it was without compare. Henry VIII had built this palace in the form of a Tudor extravaganza in 1538 near Sutton in Surrey for hunting expeditions and to entertain foreign visitors. Although built of wood, Nonsuch House was turreted like a castle, decorated with elaborate carvings and painted to look like stone. Needless to say, it became the most desirable residence on the bridge, one which only the richest merchants could afford.

Old London Bridge's days of glory nearly came to an end in 1666 when the City was almost completely destroyed by the Great Fire of London. The bridge itself survived, although several houses and the waterwheels at the northern end were burned down. Although the City was quickly rebuilt after the Great Fire, many people decided to move their residences to the attractive new squares and riverside developments that grew up in the west of London. Like the City itself, Old London Bridge ceased to be fashionable. Writing in 1750, Thomas Pennant gave a depressing description of conditions on the bridge for its inhabitants, who 'soon grew deaf to the noise of the falling waters, the clamours of the watermen or the frequent shrieks of drowning wretches'.[54]

The very structure of the bridge was also suffering. An eighteenth-century historian described it as follows:

> [There were] nineteen disproportioned arches, the starlings increased to an enormous size by frequent repairs, supporting the street above. These arches were of very different sizes and such that were low and narrow were placed between others that were broad and lofty. The back part of the houses next the Thames had neither beauty nor uniformity: the line being broken by a great number of closets that projected from the bridge and hung over the starlings.[55]

The truth is that by the eighteenth century Old London Bridge had ceased to be a wonder of the world and was looking distinctly old-fashioned and decrepit. This is shown in William Hogarth's painting of the final scene of his *Marriage à la Mode* series, in which the ramshackle houses leaning over the river form the backdrop to the death of the countess in the drawing room of her father's house overlooking the Thames.

In the eighteenth century, Old London Bridge became a byword for congestion, leading to increasing threats to build new bridges which would improve cross-river traffic by removing Old London Bridge's monopoly. The City Corporation tried to improve the situation by introducing a new rule requiring traffic to keep to the left. This decision turned out to have enormous importance in the later standardisation of the rule throughout the United Kingdom, which was formalised in the Highways Bill in 1835. It was even enshrined in a piece of doggerel:

> The rule of the road is a paradox quite,
> For if you keep to the left, you're sure to be right.

It was claimed that the 'keep left' rule was based on the fact that most people are right-handed and that it allowed them to defend themselves more easily with their swords from any approaching attacker. Unfortunately, not all countries agreed, possibly because the carrying of swords was going out of fashion, and today three-quarters of the world drives on the right.

Despite the introduction of the 'keep left' rule, Old London Bridge's monopoly was finally broken, first at Putney in 1729 and then more significantly for the City, because in central London, at Westminster in 1750. This led to serious consideration of whether the City should build a new London Bridge, or remove the houses so as to improve the flow of traffic. The City surveyor, George Dance the Elder, produced a report showing that the bridge's 500-year-old foundations were still safe and recommending that all the houses should be removed so that the roadway could be widened from 12 feet to 45 feet, and that the two central arches should be made into a single wider arch to improve navigation. In 1756, an Act was passed authorising the work at a cost of £160,000, which was financed from BHE resources.

The project started with the construction of a temporary wooden bridge that allowed traffic to cross the river while the houses were removed from Old London Bridge. On a night in April 1758, the temporary bridge burned down, causing the City to be cut off from Southwark except by ferry. There were a number of suspicious circumstances which convinced people that the fire had been caused by arson. Suspicion fell on the owners of the houses, since they had strongly objected to the destruction of their property. However, nothing was proven. The City reacted fast, hiring a thousand workmen to reconstruct the temporary bridge in less than six months. Guards were set on both bridges, and by 1762 the last house was demolished, leaving the ancient structure shorn of all its past glory.

Without its houses, Old London Bridge no longer aroused the strong emotional attachment of the past. Throughout its life, it had been in a constant state of flux resulting from changing fashions, natural deterioration and numerous fires, with the result that by now hardly any material was left from the original structure except for some underwater rubble and piling. The escalating costs of repair resulting from a series of severe winters and the continuing danger to river traffic passing under the arches led to renewed demands for a new bridge. Its death knell was sounded in 1801 by the *Third Report from the Select Committee upon the Improvement of the Port of London*, which found that Old London Bridge was now insecure. The report recommended the construction of a new bridge. Several designs were submitted, including one by the egregious Ralph Dodd. The most exciting proposal was for a 900-foot single-span iron crossing by Thomas Telford. This attractive concept was deemed impractical and no decision on a new bridge was made at the time.

In 1822, anticipating that Old London Bridge would soon be demolished, the City bought out the London Bridge Waterworks and removed the wheels. In the following year, an Act was passed 'For the Rebuilding of London Bridge and for the improving and making suitable Approaches thereto'. The City selected a design for a 1,000-foot granite structure of five semi-elliptical arches by John Rennie, whose iron Southwark Bridge had recently been completed. Sadly, John Rennie died before work could begin, and it was his son, John Rennie Jr, who took over design supervision.

The ceremony of the laying of the foundation stone took place on 15 June 1825 in the presence of the Duke of York, brother of the recently crowned George IV. A massive cofferdam was sunk 45 feet below Thames High Water so that 400 guests could descend to watch the lowering of the 4-ton lump of granite and the placement of a set of contemporary coins and an inscribed copper plate underneath. The inscription contained a eulogy to George IV as ruler of the British Empire, as well as a defence of the decision to destroy Old London Bridge. It referred to the obstruction of the free course of the river by 'the numerous piers of the ancient bridge' and stated that the City of London was 'desirous of providing a remedy for this evil by building a new, wider bridge of a character corresponding to the

William Knight's depiction of London Bridge old and new in 1832

283

dignity and importance of this royal city'. The degradation of the one-time wonder of the world was complete.

Rennie's London Bridge was constructed 180 feet upstream of Old London Bridge. The contractors, Messrs Jolliffe and Banks, completed the work on 1 August 1831. Forty workmen lost their lives during the construction project. The cost of the bridge itself was £680,232, while the cost of the approaches amounted to over £2,000,000. William IV and Queen Adelaide opened the bridge on 1 August 1831 after coming down the river in the state barge. Thousands lined the banks of the Thames, which was crowded by brightly decorated boats of all shapes and sizes.

A contemporary commentator forecast that the solid granite structure would last 1,000 years and thus prove even more long-lived than its predecessor.[56] Unfortunately, the new bridge sank one inch in each of its first eight years. By the 1960s, 50,000 commuters a day were crossing over the bridge from London Bridge Station, causing a dangerous crush on the narrow footways, while the heavy load imposed by buses and cars threatened the stability of Rennie's bridge, despite its solid appearance. Rennie's London Bridge therefore lasted *in situ* only 130 years, merely one-fifth of the life of its stone predecessor, which had been built without the aid of the mechanical and engineering advances of the Industrial Revolution. In 1967, a new London Bridge Act authorised the City Corporation to demolish Rennie's bridge and construct a new one. One of the last boats to pass under the old bridge conveyed the body of Sir Winston Churchill by river from St Paul's Cathedral following his state funeral.

The present London Bridge was designed by the firm of Mott, Hay and Anderson, with Lord Holford (1907–75) acting as architectural adviser. The contractor was the firm of John Mowlem & Co., which completed the work in 1973 for the cost of £4,500,000. Finance was provided by the BHE. The new London Bridge is 105 feet wide, consisting of three flat pre-stressed concrete arches of 260 feet, 340 feet and 260 feet. It was originally proposed that some blocks of flats should be built on the bridge so as to continue the old tradition, but this idea was

Remaining arch of Rennie's London Bridge and Nancy's steps

never likely to be accepted. As it has turned out, the plain, minimalist appearance of the current bridge could hardly be more different from the exotic idiosyncrasy of Old London Bridge. The only decorations are the granite obelisks on the faces of the piers and the polished granite facing of the parapet walls, surmounted by stainless steel handrails. Unseen by the crossing commuters is the heating system underneath the pavement which prevents the road icing over in severe winter weather, something to which Rennie's bridge was susceptible.

A remarkable epilogue to the story of London Bridge occurred with the sale of most of Rennie's stonework to the McCulloch Oil Corporation for $2,460,000, which at the time amounted to £1,000,000. McCulloch had founded a new city in the deserts of Arizona and decided that London Bridge would be a major tourist attraction. The stones were numbered and shipped to Lake Havasu City, where Rennie's bridge was re-erected. The foundation stone was laid on 23 September 1968 by the Lord Mayor of London, Sir Gilbert Inglefield, who stated in his speech that 'this

London Bridge with Fishmonger's Hall and
City skyline, viewed from the south

bridge serves as a noble and endearing monument to the strong
bonds of friendship that exist between America and England. This
foundation stone symbolises nearly 2,000 years of history and
tradition.' The reconstruction work was completed on 10 October
1971, 140 years after the original opening of Rennie's bridge. The
second opening ceremony of Rennie's London Bridge was
conducted in Lake Havasu City by another of London's Lord
Mayors, Sir Peter Studd, watched by a crowd of 100,000. The
bridge still stands there today and is approaching its double
centenary. As the traffic is light, it may yet last the predicted 1,000
years, albeit in a different climate. There is, incidentally, no truth
in the amusing story that the Americans thought they were buying
Tower Bridge rather than London Bridge.

Not all of Rennie's bridge was shipped to America. One
granite arch remains on the south bank, from which descends a
set of narrow stairs known as Nancy's Steps. This is where, in
Charles Dickens' *Oliver Twist,* Nancy has her ill-fated
conversation with Mr Brownlow in which she tells him that

Oliver is in the hands of Bill Sikes. She is overheard and this leads to her brutal murder. Dickens knew London Bridge well, as he crossed daily to Southwark to have breakfast with his father and family when they were imprisoned for debt in the Marshalsea prison in Borough High Street while he earned a living in the blacking factory at Hungerford Stairs near Charing Cross.

Remains of Old London Bridge can also be found in various other places. One of the half-domed stone alcoves erected in the 1760s to provide refuge for pedestrians now stands in the grounds of Guy's Hospital, while two more alcoves are located in Victoria Park in Hackney. The colourful coat of arms of George III which was displayed at the Southwark stone gate was bought by a Southwark publican and today can be seen in Newcomen Street. Two relics are held in Fishmongers' Hall, which stood upstream of Old London Bridge from the fourteenth century and was rebuilt in 1831 to accommodate Rennie's bridge. The first relic is a chair made out of wooden piles from Old London Bridge and decorated with carvings of that bridge, Westminster Bridge, Rennie's London Bridge and Robert Mylne's Blackfriars Bridge. Indirectly associated with Old London Bridge, the second relic is the dagger with which William Walworth stabbed Wat Tyler in 1381 before having Tyler's head impaled on the Drawbridge Gate.

Considerable remains of the second stone arch from the north were rediscovered in 1920 during excavations for the building of Adelaide House on the downstream side of London Bridge on the line of Old London Bridge. Gordon Home, author of the definitive *Old London Bridge*, tried to save the arch but failed to raise the necessary £7,000. Instead, three stones dated 1703 were raised onto the roof garden of Adelaide House. Three other stones were placed in the churchyard of St Magnus the Martyr, where they can be seen today. This location is especially appropriate since the footway onto Old London Bridge passed through here after the removal of the houses. Inside the church is a detailed model of Old London Bridge, showing the houses and hectic lives of the people who lived, worked and shopped there for over 600 years.

CHAPTER 15

Tower Bridge

Tower Bridge has now replaced London Bridge as London's most dramatic crossing. It was constructed in 1894 at the eastern end of the Pool of London, the stretch of the river downstream from London Bridge which served as London's main port from medieval times. Today, it is hard to imagine what this part of the river looked like in the nineteenth century, when ships from all over the world crowded along the wharves which lined both banks of the Thames to the east of London Bridge.

At the end of the eighteenth century, shipping here was so congested that ships would often wait for weeks to unload their cargoes. Thieving was rife, and sugar would turn to treacle before it could be landed. Parliament therefore decided to encourage the construction of a series of secure inland dock basins to the east of London by providing trade-monopoly incentives to private entrepreneurs. The nearest dock to the City, sited on the north bank of the river just to the east of the Tower of London, was St Katherine's Dock, constructed in the 1820s. This was designed by Thomas Telford, who was also working on the great suspension bridge over the Menai Straits at the same time.

Despite the extra capacity of the inland docks, the Pool of London continued to be used because of its convenient location near the centre of the City. By the second half of the nineteenth century, the British Empire had expanded to regions in all five continents, and London was the pivot of a vast system of international trade. Tea came in from China, sugar and rum from the West Indies, rice and spices from India, ivory from Africa, and grain and timber from North America. *The Royal River* describes this stretch of the Thames as it was in the reign of Queen Victoria:

> [The] wharves and warehouses [are] black with smoke of many years. Bales of goods dangle perilously as they wait to be lowered into barges which come up to landing stages. Nothing of beauty, but we see here matters of deepest interest affecting our country and her possessions in every part of the world. The greatness of Britain depends on this liberal & majestic Thames.[57]

London itself was constantly expanding eastwards. At the end of the eighteenth century, the ruins of the medieval Bermondsey Abbey could be seen in open fields on the south bank of the

William Parrott's 1840 painting of ships crowding the Pool of London

Thames between Shad Thames and Rotherhithe. However, during the nineteenth century, largely because of the construction of the docks and related industries to the east of the City, the landscape was built over to house the working population, in often crowded and unsanitary conditions. Charles Dickens describes one such area, known as Jacob's Island, where Bill Sikes lives, as 'the filthiest, strangest, most extraordinary of the many localities that are hidden in London'.

By the 1870s, a million people lived east of London Bridge on both banks of the Thames. They had no direct means of crossing the river from their homes. One option was to go further east to Wapping, where Marc and Isambard Kingdom Brunel had built their famous tunnel under the Thames in 1843. Originally, the tunnel was intended to cater for horse-drawn vehicles as well as pedestrians, but the money ran out before the necessary ramps could be constructed. Consequently, it was never financially viable, and in 1865 it was converted into a rail tunnel for the East London line, for which it is still used today. As a result, people streamed westwards in their thousands to John Rennie's rather narrow London Bridge, which was permanently clogged up with horse-drawn carts and pedestrians.

The first attempt to improve this situation was the construction in 1870 of a new tunnel under the Thames just to the west of the Tower of London. The original idea of the tunnel was to use an ambitious combination of gravity and cables to draw trams along narrow rails inside its seven-foot diameter. The system proved impractical, so instead it was always used as a pedestrian walkway. This proved so popular that a million people a year paid their halfpenny toll to pass through until it was closed in 1897. Today, the tunnel carries water mains and all that can be seen above ground is a small round access building on Tower Hill.

Despite the tunnel, public demand for a proper new bridge near the Tower of London became so overwhelming that in 1876 the City Corporation set up a committee to investigate the matter. The first problem to be solved was how to finance the bridge. Taxes and tolls were always unpopular, so it was fortunate

Tower Foot Tunnel entrance, later converted for use by the London Hydraulic Power Company

that the Bridge House Estates, originally set up to finance the building of Old London Bridge, had prospered greatly over the centuries and was now able to provide the money to build the new bridge as well as to maintain the three existing Thames bridges owned by the City Corporation – London, Blackfriars and Southwark. In order to overcome the powerful opposition of the owners of the wharves in the Pool of London, a formula for paying them compensation to cover any losses they might incur was agreed.

It remained to decide how to build a bridge in this strategic position without impeding the ships whose cargoes were so vital to London's prosperity. Over 50 designs were submitted, many of which envisaged dramatic but not always practical use of the latest technology at the time. One fantastical idea was to build a high-level bridge with enough headroom to allow ships to pass beneath, while vehicles wishing to cross over were to be raised up to it using hydraulic lifts. Another envisaged a duplex bridge with sliding road sections so that ships could pass through the open sections while vehicles used the closed ones.

Sir Joseph Bazalgette, chief engineer of the Metropolitan Board of Works, submitted several designs, including one with a gigantic parabolic steel arch from which was hung a high-level roadway. This design bore a close resemblance to the Austerlitz Viaduct which was built at about the same time across the River Seine in Paris. In the end, the competition was narrowed down to the technologically modern designs by Bazalgette and a design for a medieval-looking drawbridge by the City Corporation's architect, Horace Jones. At first sight, it must surely have seemed that this would be a one-sided contest. The Industrial Revolution in Britain had seen engineers eclipse architects in the design of so many pioneering structures, such as railways, train sheds and bridges, throughout the country, and Bazalgette was by now the pre-eminent engineer, following his success in building London's sewers and embanking the Thames. However, Jones did have one advantage, in that as chief City Architect he was asked by the committee to comment on Bazalgette's designs. Despite this conflict of interests, his criticisms were probably fair, and the committee decided that none of Bazalgette's proposals provided sufficient headroom for tall ships to pass through and, besides, the approach roads would have been too steep for horse-drawn vehicles. Consequently, Jones was selected to design the new Tower Bridge.

Horace Jones (1819–87)

Horace Jones was born in London. He became apprenticed to William Tite, the architect of the Royal Exchange, and when Jones set up his own practice in 1842, Tite remained a strong supporter of his often controversial work. Jones's massive Cardiff Town Hall of 1854 was opened to a mixture of acclaim and criticism which presaged the conflicting reactions which were to greet the opening of Tower Bridge 40 years later. When the post of City Architect came up for election in 1868, Tite commended Jones's application for his 'diligent, earnest and attentive' approach, while Sir John Carmichael, one of his clients, commended him for 'completing his work with strict regard to the financial portion of the contract'. No doubt these qualities appealed to the City fathers who appointed him to the post. He proceeded to design many public and commercial buildings, including the massive Smithfield Market, which has been described as a 'cathedral to meat', and Billingsgate Fish Market, which fronts the north bank of the Thames 400 yards upstream of today's Tower Bridge. Horace Jones became president of RIBA from 1882 to 1884 and was knighted in 1886.

Despite being trained in classical architecture, Jones had become a proponent of the Gothic revival style, as epitomised by the Houses of Parliament, designed by Sir Charles Barry and Augustus Pugin. This fitted in well with the City Corporation's stated requirement of harmonising the appearance of the new bridge with the Norman architecture of the nearby Tower of London, the city's most historic building. However, Jones had no experience of bridges, so he was asked to collaborate with the engineer John Wolfe Barry, who by an amazing stroke of fate was the son of Jones's architectural mentor, Sir Charles Barry.

John Wolfe Barry (1836–1918)

John Wolfe Barry was the youngest of five brothers, one of whom was the eminent architect E.M. Barry. He was apprenticed to Sir John Hawkshaw, under whom he worked on Charing Cross and Cannon Street railway bridges. In 1867, he formed his own engineering firm and undertook a wide variety of successful projects. These included the Barry Docks in Cardiff, extensions to the District Underground line and two bridges over the Thames – Kew Road Bridge and Blackfriars Railway

Bridge. His experience in laying towers in soft subsoil under water was to prove invaluable in the construction of Tower Bridge. Barry later became the president of the Institution of Civil Engineers and was knighted in 1897 after the completion of Tower Bridge. His obituary in the January 1918 edition of *Engineering* praised his 'wide experience, sound judgement and untiring energy, which made him the leader of his profession in Great Britain'.

The collaboration between Jones and Barry proved fruitful. Jones's original sketch of a bascule bridge had appealed to the Corporation because it solved the problem of allowing access to shipping while conforming to the Gothic architecture of the eleventh-century Tower of London. However, the method of raising the bascules using steam-powered hoisting machinery was impractical, and the arched structure of the central span inhibited raising of the bascules to their fullest extent, which would impede the passage of tall sailing ships. Barry made the necessary changes, and the design submitted to Parliament in 1884 closely resembles the bridge we see today.

After much discussion, a parliamentary Act was passed in August 1885 'to empower the Corporation of London to construct a bridge over the River Thames near the Tower of London, with approaches thereto'. The final design as we know it today provided for a low-level road bridge which can be raised at the centre, together with two high-level walkways. This combination allowed even the tallest ships to pass through while pedestrians could still cross over using the walkways, which were accessible via steps or lifts.

The design encompasses four towers. The taller central towers provide support for the walkways, which rise 120 feet above road level and allow headroom of 135 feet for ships to pass underneath at high tide. The shorter bank-side towers support the chains used to suspend the road platforms which span the river as far as the central towers. At the top of the central towers,

Tower Bridge's high-level
footways

the chains are joined by tie-rods, which are cunningly concealed in the wrought-iron parapets of the two footbridges. All four towers are of granite and Portland stone, designed in a Gothic Scottish-baronial style in order to harmonise with the architecture of the Tower of London. However, all is not what it seems. In fact, each of the towers is constructed of four octagonal steel columns behind the cladding of stone. As an engineer, Barry may have been worried by the requirement that the real nature of the bridge was disguised. However, he justified this by pointing out that iron and steel are best protected from corrosion by encasing them in masonry. In a lecture in 1894, he admitted that 'some purists will say that the lamp of truth has been sadly neglected in this combination of materials', but he hoped that 'we may forget that the towers have skeletons as much concealed as that of the human body, of which we do not think when we contemplate examples of manly or feminine beauty'.[58]

Construction of Tower Bridge started in 1886, and the Memorial Stone was laid by the Prince of Wales on 21 June. Commenting on the ceremony, the 26 July 1886 edition of *The Graphic* welcomed the bridge as a free gift to Londoners from the Bridge House Estates. It also attacked the vested interests which had so long opposed a new bridge here and opined that the bridge might never have been built if Parliament's powers had not been augmented by the latest Reform Act. In a final prophetic comment, it suggested that before long a new bridge even further to the east would be needed as London by then began at Tilbury.

During the ceremony, Horace Jones was nearly killed by falling machinery, but he survived to receive a knighthood in August of that year. Sadly, he died of heart disease in 1887 and so never saw the completion of his most famous work. It was up to Barry to see the project through. He made a number of changes to the design of the bridge, including adding the shorter abutment towers which had not appeared in Jones's original design. One highly significant change was also made to the appearance of the bridge. Jones had originally envisaged cladding the steel of the towers in red brick, but his assistant and successor George Stevenson insisted on using stone. It is fascinating to speculate what Tower Bridge would have looked like if Jones had survived longer.

The construction project provided an enormous challenge. J.E. Tuit, the chief engineer for Sir William Arrol & Co., commented:

> how difficult a problem the crossing of a river may be when the banks on one side of that river are very low, the river full of shipping, and the vested interests in the wharves on each side very large.[59]

This was written in 1894, after the bridge was finally opened four years later than originally estimated. One major cause of the slippage was the insistence by the City Corporation, under pressure from the wharfingers, that a width of 160 feet of clear water be maintained in the middle of the river throughout the project so that ships could pass unhindered. This had a major effect on the pace of construction of the foundations of the tall river-towers, as it meant that they could not be worked on simultaneously.

It was decided to build the foundations of each of the river-towers in 12 caissons sunk 21 feet into the river-bed. Each caisson was an open-ended box of wrought iron which was lowered from a wooden platform. Divers were first sent down to dig into the gravel and clay by hand so that the caissons sank inch by inch to an initial depth of eight feet. They worked in teams of four in nine-hour shifts. On average, they managed to lower a caisson eight inches per day into the river-bed. After this slow and dangerous process, the water could be pumped out of the caissons so that navvies could descend and dig out the space more efficiently until the bottom of the caisson had reached the required depth of 21 feet. Finally, the caissons could be filled with concrete to form the foundations of the river-towers.

The men who worked on this stage of the project, the divers and navvies, were almost a race apart from the rest of the population. Diving was a highly specialised and dangerous occupation in the nineteenth century, as indeed it is today, even with much safer modern equipment. The first satisfactory breathing device was invented in the 1820s by John Deane of Whitstable after he had rescued horses trapped in burning stables by using a helmet from an old suit of armour and connecting it to an air line and pump so that the farmer could feed him with air while he led the horses out of the smoke-filled stables. Later, a Prussian engineer, Augustus Siebe, improved the design to produce a combined helmet and diving suit which was the basis for all diving equipment until well into the twentieth century. Victorian divers formed an elite band whose skills were most frequently used for salvage work on sunken ships, and tended to live a life apart. They had a reputation for being able to out-drink even the navvies.

The name 'navvy' comes from the men who worked on the navigation canals at the end of the eighteenth century and was taken over by the workers who performed a similar role on the construction of the railways from the 1820s onwards. Many were Irish, but there were also Scottish and English as well as some foreign navvies. In the early days, they formed an anarchic group of labourers moving from place to place to work on the vast number of railways and bridges that covered the country. Thomas Carlyle wrote about this period of hectic construction: 'All the world calculates on getting to heaven by steam. I have not on my travels seen anything uglier than the disorganised mass of labourers, sunk three-fold deeper in brutality by the three-fold wages they are getting.' Certainly, they were relatively well paid, as the contractors had to give them an incentive to complete often dangerous tasks on time. However, they were notorious for wild drinking sprees after receiving their weekly wages and often lost at least a day's paid work while recovering from the after-effects. By the 1860s, most of the great railway projects had finished and the remaining navvies, such as the men who were employed on digging the foundations for Tower Bridge, lived a less anarchic life.

The contractors selected to construct the various stages of the bridge were among Britain's and the world's most renowned and experienced at the time. The contractor for all the steel work was Sir William Arrol & Co., which was responsible for the structural steel columns supporting all four towers, as well as the lattice girders of the high-level walkways. As much as possible was prefabricated at Arrol's Dalmarnock works near Glasgow. However, the installation of the massive steel frames and the riveting of the walkway girders required hundreds of men to work at increasingly high levels above the river. It says much for Arrol's efficiency and concern for the welfare of his workers that not a single death occurred during this phase of the project. The especially dangerous work of fixing the walkways was done from a wooden platform attached underneath. This method had the additional safety benefit that it prevented any tools or construction material falling onto the ships passing below. Any damage caused to shipping would have had a disastrous effect on the project.

William Arrol (1839–1913)

Born in Paisley, Scotland, William Arrol received no formal education. He served his apprenticeship as a blacksmith before getting a job at Laidlaw's engineering works in Glasgow, where he soon became foreman. In 1868, he started his first business as a boilermaker and in 1872 built the Dalmarnock Ironworks, also near Glasgow, which was to become the largest structural steelworks in the United Kingdom, employing 5,000 workers. He earned a reputation as a model employer with a special concern for quality of work and the safety of his workers. One of the first major projects he undertook was the construction of the new bridge over the River Tay after the disaster of the destruction of the first bridge on the stormy night of 28 December 1879. On that night, the 13 central spans had been blown down, taking with them a train and 75 passengers and crew, who all perished in the Firth of Tay in one of the greatest disasters that has ever happened to an engineering structure. The Court of Inquiry found that the bridge design and quality of construction were inadequate to withstand the storm-force winds that can occur in the area. These findings had a profound effect on the future of bridge design, as well as probably leading to the premature death of the designer, Thomas Bouch, who had just been knighted for the construction of the seemingly impressive bridge. Arrol is most famous for his next major project, the construction of the Forth Railway Bridge. One of the first bridges to be made of steel, this mighty structure, consisting of three immense cantilevers and connecting girder spans, stretches to a total length of 1.5 miles and is still one of the largest bridges of its kind in the world. For this achievement, Arrol received a knighthood, and he was the obvious choice for building the superstructure of Tower Bridge.

The raising of Tower Bridge's bascules

Tower Bridge is probably best known for its impressive central drawbridge. Despite its distinctly medieval appearance, the drawbridge was built to be operated by the most advanced technology available in the nineteenth century – hydraulic power. Driven by steam pumping engines, the system was installed by Sir William Armstrong (1810–1900), the pioneer of hydraulic machinery, whose inventions were widely used for lifting equipment in docks, mines and railways and even for raising the curtains in London theatres. He later became famous as the manufacturer of the Armstrong breech-loading gun, and after his death the output of this weaponry from the amalgamated firm of Vickers Armstrong helped win the Second World War. The drawbridge itself consists of two 1,200-ton bascules which can be raised to their fullest extent of 80 degrees in an amazing 60 seconds, allowing a ship to pass through in five minutes, by which time the bascules can be lowered and road traffic resumed.

The bascules work on the see-saw principle (*bascule* is the French word for see-saw). Each bascule is counterbalanced by a

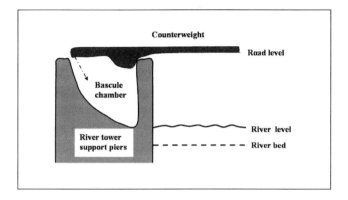

The operation of the bascules

semicircular-shaped iron-and-lead weight of 422 tons, which is housed in a cavernous chamber in the top of the supporting piers of each of the river-towers, together with the machinery to raise the bascule. It is an awesome experience to stand at the bottom of this chamber and look up at the massive structure of the bascule and its counterweight towering above and ready to descend if the bridge is to be raised.

As was common with major construction projects in those times, Tower Bridge was not completed without loss of life. According to the resident engineer, George Crutwell, four men died during the dangerous task of laying the foundations deep in the river. In addition, one man died during the construction of the approach roads and five working on the superstructure, making a total of ten. Barry himself records only six deaths. There were no health and safety regulations at the time, no safety harnesses, and the workmen wore soft caps rather than the hard hats required on today's building sites. For an 8-year project with an average of 432 men employed throughout, it is certainly remarkable that there were so few fatalities. This compares with 57 fatalities during the much larger Forth Bridge project, which William Arrol had worked on just prior to Tower Bridge.

Tower Bridge was finally opened on 30 June 1894 by the Prince of Wales. Massive crowds lined both sides of the Thames, and as many as could watched from boats as the royal party approached. *The Times* recorded the most dramatic moment of all:

> The Prince turned the lever of the valve communicating with the hydraulic machinery and straightaway the two ponderous leaves, each 115 ft long, began as if by magic to rear themselves into the air. Forthwith the blare of trumpets was drowned in the wild whistling of the steamers and in the shouts of the people without. Few sights more imposing and majestic have ever been seen in this country than the silent irresistible upheaval of these solid leaves of the bridge which fascinated the spectators on land and water.

The project had taken four years longer than originally planned and the total cost was nearly £1,200,000, compared with Barry's original estimate of £750,000. Press comments at the time were by no means all favourable. Some thought the extra expense of providing a see-saw bridge was unnecessary and others complained about the use of stone to disguise the real nature of the structure. However, the public soon took it to their hearts, and, according to Honor Godfrey, a visiting Frenchman was highly complimentary about the bridge and, perhaps more surprisingly, about the English national character it seemed to exemplify.[60]

During its first year of operation, Tower Bridge was opened on average 17 times a day. Ships always had priority over land traffic and could request that it be opened as they approached, using semaphore in daytime, lamps at night and gongs during the frequent fogs that plagued London. This system lasted until 1962, by which time river traffic was much reduced, and today a ship must give 24 hours' notice if it wants to pass through. It had been expected that people would ascend to the high-level walkways, either by hydraulic powered lifts or by climbing the 206 steps to the top, to avoid waiting for the ships to pass. However, most preferred to wait and watch the passing ships, and so little use was made of the walkways. Crime was also a problem. Prostitutes plied their trade and there were frequent muggings. For all these reasons, the walkways were closed from 1910. There is no truth in the story that the walkways were closed because it was a popular jumping-off platform for attempted suicides. In fact, the lattice girders along the sides of the walkways make it difficult to jump down, and there is no record of a suicide attempt from the walkways, although one man did jump for a bet and disappeared never to be seen again.

It is, however, true that the sinisterly named Dead Man's Hole, situated by the river under the north approach road, was in the past used to retrieve and store bodies found in the river near by until they could be removed for burial. Some of these may have been suicides who jumped from the road bridge. Following the *Marchioness* disaster of 1989, several inquiries were held into river safety. Eventually, in 2002, coordination of all rescue operations was handed over to the Maritime and Coastguard Agency (MCA) and four Royal National Lifeboat Institution (RNLI) stations were set up on the Thames. According to MCA statistics, 28 people lost their lives in the Thames in 2004–5, compared with an average of 40 to 50 in the past. About 80 per cent of these deaths are by suicide, mainly by jumping off one of the Thames bridges. A larger number are rescued from the river.

Historically, Waterloo Bridge was known as the main suicide bridge, but today more attempts are made from Tower Bridge. It is thought that one of the causes of the reduction in suicides is the prevalence of mobile phones, as the police now often receive warnings when a bystander sees a person preparing to jump. Today, if a body is recovered from the Thames, it is taken for identification to the mortuary at Wapping Police Station, near the Thames River Police blue-and-white boathouse, which can be seen on the north bank downstream from the bridge. Most can now be properly identified by use of modern technology, including links to Scotland Yard's database of missing persons and from DNA, fingerprints or dental records. It is often difficult to establish whether or not death was actually a result of suicide, and in this case an open verdict or a verdict of death by misadventure is recorded. Only if there is clear evidence of intent, such as a letter or the discovery of heavy stones in the person's clothes, is a verdict of suicide recorded.

During its life, Tower Bridge has changed colour several times. Originally, it was chocolate brown, then battleship grey, before being painted red, white and blue in 1976 in preparation for the Queen's Silver Jubilee in 1977. Today's light-blue-and-white decoration is therefore not historically correct and may not be the final incarnation.

Several daring exploits are connected with Tower Bridge. In 1912, Frank McClean became the first person to fly under the walkways and between its towers in his Short pusher biplane. He carried on up the Thames and back again, but the police refused to allow him to repeat the performance on his return journey. On 28 December 1952, a number 78 double-decker bus was travelling across the bridge when the northern bascule started to rise. The driver, Mr Albert Gunter from Islington, made an instant decision to accelerate so that the bus successfully jumped over the gap. As reported in *The Times*, he said: 'I had to keep going otherwise we would have been in the water. I suddenly saw the road in front of me appeared to be sinking. In fact the bus was being lifted by one half of the bridge.' Ten people were injured as a result: the driver, the conductor and eight passengers. The experience resulted in Mr Gunter having a nervous breakdown, and for this he received compensation. Normal practice was that a red light should have been on to warn of the raising of the bascules and a hand bell should have been rung. It seems that a relief man was operating the bridge at the time and it is possible that the driver did not receive proper warning. Only one other similar incident has been recorded, when in 1943 a driver did not see the red light, tried to stop when he saw the south bascule rising but skidded into it. Fortunately, no great damage was done.

During the Second World War, despite massive bombing raids on the City of London and the docks nearby, Tower Bridge was left virtually intact and, like St Paul's Cathedral, stood proud amongst the surrounding destruction. Evidently the Luftwaffe decided that the bridge would serve as an excellent navigation aid and so did not want to destroy it. It is indeed amazing that not one of London's river bridges was destroyed during the war. The reason, according to Tower Bridge engineers, was that Hitler was so confident of victory that he decided to leave them undamaged for use by his invading forces. Sadly, one of the bridge's service tugs was hit by a doodlebug in 1944, and it sank with all its crew.

Having survived the war, albeit with some minor damage, Tower Bridge was subjected to a further threat – that of post-war architectural whims. It was clear that the drawbridge was now no longer essential, as far fewer ships needed access to the Pool of London. The architect W.F.C. Holden therefore proposed cladding the whole of the bridge in a massive glass and steel canopy so as to provide space for shops and offices, while removing the bridge's most exciting feature, the moving bascules. Fortunately, this idea was rejected and the bridge was restored with the bascules intact.

Since then, there have been many ceremonial openings of the drawbridge, including the return of the royal yacht *Britannia* from the Queen's world tour of 1953–4 and the completion of Sir Francis Chichester's solo voyage round the world in his diminutive yacht *Gipsy Moth IV*, which is today on view in Greenwich. Booked openings still give ships priority over road traffic. This nearly led to an international incident when President Clinton and his wife, Hillary, were returning from being entertained to lunch at the Pont de la Tour restaurant, from which they would have had an impressive view of Tower Bridge from the south bank. To the President's chagrin, when his car approached the bridge a ship was about to pass through and his party had to wait until the bascules had been raised and then lowered.

In 1976, the steam-pumping engines which controlled the bascule-raising machinery were finally replaced by electric engines, marking the end of an era for hydraulic power. In 1982, the high-level walkways were reopened as part of the Tower Bridge Exhibition, where the still magnificent old pumping

The steam engines inside
Tower Bridge's engine house

engines can also be seen in full steam, although they no longer
drive any machinery.

Today, Tower Bridge stands as a memorial to the wonders of
Victorian engineering. However, it would certainly be a very
different bridge if it were built today. Barry himself said in his
lecture of 1894: 'as road traffic increases and river traffic goes
more into the docks the fate of the bridge will be to become
fixed'. This seemingly modest and far-sighted prediction was
wrong in two respects. First, river traffic has decreased to such an
extent that neither the docks nor the wharves exist today.
Telford's St Katherine's Dock, just to the east of Tower Bridge on
the north bank, has been converted into a marina. The great
wharves on the south bank have been converted into shops and
restaurants, and the Anchor Brewery which operated on the
south bank just to the east of the bridge until 1980 has been
converted into penthouses. Second, his bridge still moves. The
bascules are opened about 1,000 times a year and provide one of

the most exciting views in London. Tower Bridge has become a major tourist attraction. It is London's most instantly recognisable landmark and as much a symbol of the city as the Leaning Tower is of Pisa or the Eiffel Tower is of Paris.

Many other impressive buildings front the river here, including the Tower of London, which is seven centuries older, and City Hall, which is a century younger, but there is no doubt that Tower Bridge dominates its surroundings as a major landmark and the eastern river gateway to London.

Thames Bridges Summary

Albert Bridge (1873)
710 feet long, 40 feet wide
Twin ornamental cast-iron towers resting on concrete foundations support the carriageway by cable-stayed rods, which fan out from the top of the towers, and by suspension chains. In the 1970s, Albert Bridge had to be strengthened by the installation of two cylindrical concrete river-piers to support the carriageway.
Engineer: R.M. Ordish. Contractor: Messrs Williamson & Co.

Barnes Railway Bridge (1895)
Replaced Joseph Locke's cast-iron railway bridge of 1849.
360 feet long
Three spans of wrought-iron bowstring girders carry two railway tracks across the river. Locke's original structure still stands, unused, on the upstream side.
Engineers: London and South Western Railway. Contractor: Head, Wrightson & Co.

Battersea Bridge (1890)
Replaced Henry Holland's wooden bridge of 1771, immortalised in Whistler's painting *Nocturne: Blue and Gold.*

670 feet long, 55 feet wide

Five spans, each consisting of seven cast-iron arched ribs, support the 40-foot-wide roadway and two footpaths, which are cantilevered out from the main structure. Ornamental shields in the spandrels and Moorish-style arches on the parapet enhance the bridge's appearance.

Engineer: Joseph Bazalgette

Battersea Railway Bridge (1863)
670 feet long

Five iron arched spans are supported by four stone-faced river-piers.

Engineer: William Baker

Blackfriars Bridge (1869)
Replaced Robert Mylne's elegant stone bridge of 1769.

963 feet long, 105 feet wide

Five wrought-iron spans rest on massive river-piers ornamented with red polished-granite columns. The capitals of the columns are carved with interlaced birds and plants, and support pedestrian refuges. Widened from 75 feet to 105 feet in 1909.

Engineer: Joseph Cubitt. Contractor: Messrs P. & A. Thorn & Co.

Blackfriars Railway Bridge (1886)
933 feet long

Five spans of wrought-iron arched ribs support the railroad, which provides seven tracks.

Engineers: John Wolfe Barry and H.M. Brunel. Contractor: Messrs Lucas & Aird

Cannon Street Railway Bridge (1866)
855 feet long

Five utilitarian spans of wrought-iron plate girders supported by cast-iron columns carry ten rail tracks across the river to the once magnificent Victorian train shed of Cannon Street Station.

Engineer: John Hawkshaw. Contractor: South Eastern Railway

Chelsea Bridge (1937)
Replaced Thomas Page's much-admired suspension bridge of 1858.

698 feet long, 83 feet wide

Two 55-foot-tall plain, square towers support the suspension chains from which the roadway is hung. At either end, lamp-posts decorated with golden galleons relieve the otherwise unexciting design.

Architects: G. Topham Forrest and E.P. Wheeler. Engineers: Rendel, Palmer and Triton. Contractor: Messrs Holloway Bros (London) Ltd

Chiswick Bridge (1933)
450 feet long, 70 feet wide

Three flat ferro-concrete arches are faced with Portland stone.

Architect: Herbert Baker. Engineer: Alfred Dryland. Contractor: Cleveland Bridge & Engineering Co. Ltd

Golden Jubilee Bridge (2002)
325 metres long, 4.7 metres wide

Two footbridges on either side of Hungerford Railway Bridge are supported by white-painted steel rods which fan out from slanting steel pylons.

Architect: Lifschutz Davidson. Engineer: WSP Group. Contractor: Costain/Norwest Holst

Grosvenor Railway Bridge (1860)
700 feet long

Originally, the four wrought-iron spans carried four rail tracks across the river. In 1965, the bridge was reconstructed in steel and now provides a crossing for ten tracks. In fact, the steel structure consists of ten separate bridges joined together.

Engineer: John Fowler

Hammersmith Bridge (1887)
Replaced Tierney Clark's elegant structure of 1827, which was the first suspension bridge to cross the Thames.

688 feet long, 33 feet wide

Two river-towers of wrought iron clad in highly ornamental cast iron support steel suspension chains from which the narrow carriageway is hung. The footways are cantilevered out from the main structure.

Engineer: Joseph Bazalgette. Contractor: Messrs Dixon, Appleby and Thorne

Hungerford Railway Bridge (1864)
Replaced Brunel's suspension footbridge, the chains of which were removed for use in the Clifton Suspension Bridge.

1,200 feet long

Nine wrought-iron girders are supported on cast-iron cylinders and on the two arched brick river-piers preserved from Brunel's suspension bridge. The bridge was widened in 1886 to increase the number of railway tracks from four to eight.

Engineer: John Hawkshaw. Contractor: South Eastern Railway

Kew Bridge (1903)
Replaced Robert Tunstall's wooden bridge of 1759 and James Paine's stone bridge of 1789.

360 feet long, 56 feet wide

Three rough granite elliptical arches are enhanced by the ornamental shields of the counties of Middlesex and Surrey carved into the walls.

Engineer: John Wolfe Barry. Contractor: Easton Gibbs

Kew Railway Bridge (1869)
575 feet long

Five wrought-iron lattice-girder spans supported on cast-iron columns with ornate capitals carry two railway tracks across the river.

Engineer: W.R. Galbraith

Lambeth Bridge (1932)

Replaced P.W. Barlow's suspension bridge of 1862.

776 feet long, 60 feet wide

The five arches of the bridge, supported by granite-faced river-piers, are faced with flat steel plating to disguise the steel skeleton that lies behind. The red colour scheme is intended to reflect the red furnishings of the nearby House of Lords. Pineapple obelisks stand at the approaches.

Architect: Reginald Blomfield. Engineer: George W. Humphreys. Contractor: Dorman, Long & Co. Ltd

London Bridge (1973)

Replaced John Rennie's London Bridge of 1831, which itself replaced the inhabited Old London Bridge of 1209.

860 feet long, 105 feet wide

Three cantilevered high-strength concrete arches have spans of 260 feet, 340 feet and 260 feet. The only decorative features are the granite obelisks on the river-piers and the polished-granite facing of the parapet walls.

Architect: Lord Holford. Engineers: Mott, Hay and Anderson. Contractor: John Mowlem & Co.

London, Chatham and Dover Railway Bridge (1864)

933 feet long

Five spans of wrought-iron lattice girders were supported by massive cast-iron columns. The superstructure was removed in 1985, leaving just the headless columns.

Engineer: Joseph Cubitt. Contractor: Kennards of Monmouthshire

Millennium Bridge (2002)

325 metres long, 4.7 metres wide

The flat steel suspension bridge carries pedestrians over the river between Tate Modern and St Paul's Cathedral. Also known as the 'Wobbly Bridge' because of the swaying that occurred at its official opening in 2000. The bridge was closed while the problem was solved using a system of dampers.

Architect: Foster and Partners. Engineer: Ove Arup & Partners. Contractors: Monberg & Thorsen/Sir Robert McAlpine

Putney Bridge (1886)
Replaced the wooden Fulham Bridge of 1729.
   700 feet long, 74 feet wide
   Five segmental granite arches span the river with All Saints Church, Fulham, at the northern end and St Mary's Church, Putney, at the southern end.
   Engineer: Joseph Bazalgette. Contractor: John Waddell

Putney Railway Bridge (1889)
750 feet long
   Five turquoise wrought-iron lattice girders supported by pairs of cast-iron cylinders provide two railway tracks for the District Line.
   Engineer: William Jacomb. Contractor: Head, Wrightson & Co.

Richmond Bridge (1777)
280 feet long, 36 feet wide
   Five segmental arches are constructed in masonry faced with Portland stone. Widened on the upstream side in 1939.
   Architect: James Paine. Engineer: Kenton Couse. Contractor: Thomas Kerr

Richmond Footbridge, Lock and Weir (1894)
300 feet long, 28 feet wide
   Twin high-level footbridges pass over a lock capable of handling six river barges and a weir controlled by lifting sluice gates. Originally hand cranked, the sluice gates are now raised by electric power.
   Engineer: F.G.M. Stoney. Contractors: Ransomes and Rapier

Richmond Railway Bridge (1848)
300 feet long
   Three 100-foot steel girders are supported on stone-faced land arches and two stone-faced river-piers. The original cast-iron girders were replaced by steel in 1907.

Engineer: Joseph Locke

Southwark Bridge (1921)
Replaced John Rennie's three-span iron bridge of 1819.
   800 feet long, 55 feet wide
   Five steel arches are supported by four stone river-piers, which are topped by pierced lunettes for decoration.
   Architect: Ernest George. Engineer: Mott, Hay and Anderson. Contractor: Sir William Arrol & Co.

Tower Bridge (1894)
880 feet long, 60 feet wide
   Central drawbridge with two bascules of 1,100 tons each, originally raised by steam-driven hydraulic power, today by electricity. Two 300-foot steel towers clad in granite and Portland stone support the bascules as well as a 200-foot-high walkway which is cantilevered out from the towers. Suspension chains support the road spans from the riverbanks to the two towers.
   Architect: Horace Jones. Engineer: John Wolfe Barry. Contractors: Sir William Arrol & Co. and William Armstrong

Twickenham Bridge (1933)
280 feet long, 70 feet wide
   Three reinforced-concrete arches are supported on concrete river-piers, with bronze plated permanent hinges at the springings and centres to allow adjustments due to changes in temperature.
   Architect: Maxwell Ayrton. Engineer: Alfred Dryland. Contractor: Aubrey Wilson Ltd

Vauxhall Bridge (1906)
Replaced James Walker's bridge of 1816, which was the first iron bridge to be built over the Thames in London.
   759 feet long, 80 feet wide
   The appearance of the structure of five steel arches is enlivened by the heroic-sized statues which stand in front of each of the river-piers.

Architect: W.E. Riley. Engineers: Alexander Binnie and Maurice Fitzmaurice. Contractor: Petwick Bros

Wandsworth Bridge (1940)
Replaced J.H. Tolmé's five wrought-iron arches of 1873.
619 feet long, 60 feet wide
Three steel cantilever spans are supported by granite-faced river-piers in a typically plain LCC design.
Architect: E.P. Wheeler. Engineer: T. Pierson Frank. Contractor: Messrs Holloway Bros (London) Ltd

Waterloo Bridge (1945)
Replaced John Rennie's 1817 bridge of nine semi-elliptical granite arches, which was once described by Canova as 'the noblest bridge in the world'.
1,200 feet long, 80 feet wide
Five spans of reinforced concrete clad in Portland stone cross the river between the modernist concrete structures of the South Bank Centre and the classical stone structure of Somerset House on the north bank. Externally, the spans appear as elegantly flat arches, but the underlying structure consists of steel box-girders.
Architect: Giles Gilbert Scott. Engineer: Rendel, Palmer and Triton. Contractor: Sir William Arrol & Co.

Westminster Bridge (1862)
Replaced Labelye's beautiful but unsafe stone bridge of 1750.
748 feet long, 85 feet wide
Seven elliptical cast- and wrought-iron arches supported by granite piers cross the river between the former County Hall and the Houses of Parliament. Gothic shields in the spandrels and ornamental shields emblazoned with the arms of England and Westminster provide decoration appropriate to the site.
Architect: Charles Barry. Engineer: Thomas Page. Contractor: Thomas Page

Table of daily vehicle crossings over London's Thames bridges

Bridge	Crossings/ 24 hours	A road
Richmond Bridge	34,484	A306
Twickenham Bridge	46,188*	A219
Kew Bridge	41,561*	A217
Chiswick Bridge	39,710*	A3220
Hammersmith Bridge	24,203	A3031
Putney Bridge	58,687	A3216
Wandsworth Bridge	53,299	A202
Battersea Bridge	26,041	A3202
Albert Bridge	19,821	A23
Chelsea Bridge	29,375	A301
Vauxhall Bridge	50,533	A201
Lambeth Bridge	25,187	A300
Westminster Bridge	32,673	A3
Waterloo Bridge	41,960	A100
Blackfriars Bridge	38,982	A201
Southwark Bridge	12,465	n/a
London Bridge	35,345	A10
Tower Bridge	40,024	A100

The daily numbers of vehicle crossings were measured in 2004 after the introduction of the congestion charge, except for the asterisked figures.

Figures supplied by Transport for London.

APPENDIX 2

Bridge Basics

The design and construction of a bridge are highly technical exercises. This book can give only a general description of some of the basic concepts relating to the different kinds of bridges and the materials used to construct them. In summary, there are six main types of bridge design to be found on the Thames. These are beam, cantilever, arch, suspension, cable stay and bascule. The situation is often more complex, when several of these types are combined in a single bridge, as for example in Tower Bridge.

Having decided on what type of structure to design, the engineer needs to consider the choice of materials. Wood, stone, cast and wrought iron, steel, concrete and reinforced concrete have all been used in various combinations. These basic design choices affect the most important aspects of a bridge, including the width of the navigable spans, the road width, the load it can carry, its ability to withstand the impact of weather and tide, and its aesthetic appearance.

The most basic type of bridge is the beam. In its simplest form, this would consist of a log of wood thrown over a stream and supported by the banks on either side. The Thames is too wide to allow for such a simple structure, and the spans of its beam bridges are supported by several river-piers, as well as piers on

317

the riverbanks. All the Thames railway bridges are designed as beams, usually with steel lattice girders crossing the river supported by iron or steel piers.

The cantilever often looks like a beam, but its weight is not supported at both ends by lying across two piers, as is the case with a beam. In fact, one end is held down by a bracket or heavy weight at a supporting pier or abutment while the other end is built out from there without the need for any further support. Two cantilevers opposite one another can be joined together to form the simplest type of cantilever bridge, but in practice several cantilever spans are needed to cross the Thames. The walkway at the top of Tower Bridge is an example of a simple cantilever construction, even though it looks like a single beam.

The main problem with the beam and cantilever before the age of iron and steel was that stone does not have enough tensile strength to allow for long spans. This means that the tensile, or stretching, effect on the underside of a long stone beam causes it to break under the combined weight of its own load and the load of the traffic passing over. On the other hand, stone does have great compressive strength. This means that it can withstand considerable inward or downward pressure. The arch was used by the Romans to build stone bridges that have lasted for two millennia.

The arch is built by constructing a wooden framework, or centering, in the shape of an arch between two piers. Then cut stones, known as voussoirs, are laid across the centering and the keystone is inserted in the middle of them. The wooden framework is then removed, and the weight of the stones and any loads on top of them simply presses down towards the ground, thus utilising the considerable compressive strength of stone and avoiding its tensile weakness. Early stone arched bridges over the Thames, such as Richmond Bridge, used semicircular arches in which the direction of the pressure is mainly downwards. This limited the width of the spans. Later bridges, such as Mylne's Blackfriars Bridge, used elliptical arches with wider spans. Here the pressure is outwards as well as downwards, and this requires strong buttresses at the abutments to control the outward thrust.

A suspension bridge consists of a roadway which is supported from above by vertical cables. The cables in turn hang from long curved chains, or catenaries, which are supported on tall river-towers and tied down in abutments on the riverbanks. This design provides a wide navigable river span and looks dramatic. The disadvantage lies in the complex calculations needed to ensure that the many and various tensile and compressive forces are catered for, to ensure that traffic loads and high winds can be withstood.

A cable-stayed bridge such as the Albert Bridge looks at first sight like a suspension bridge. In fact, the roadway is supported by straight cables hanging directly from the river-towers without any catenaries.

A bascule bridge such as Tower Bridge consists of two cantilevers counterbalanced by heavy weights. The cantilevers can be moved up and down mechanically like a drawbridge to allow shipping to pass underneath. The operation is described in more detail in Chapter 15 on Tower Bridge (see the diagram on p. 301).

Regarding the choice of materials, the characteristics of stone and its ability to tolerate compression but not tension are described above. Some early bridges, such as Fulham Bridge, were made of wooden beams. Wood has rather more tensile strength than stone – it can bend as it is stretched under a load. However, in every other way it is less durable, and it ceased to be used after the middle of the eighteenth century.

As for iron, two main sorts are used in bridges: cast iron, which has a high carbon content and is relatively brittle; and wrought iron, which contains hardly any carbon and is more malleable. The first iron bridge, at Coalbrookdale over the River Severn, was built of cast iron using an arched construction. Since the main stress with an arch is downwards compression, this bridge has stood the test of time. Cast iron has also been successfully used for the river-piers of several Thames bridges, where again compressive strength is required. However, because it is extremely brittle, cast-iron beams such as lattice girders tend to break as they are stretched under heavy loads over time. Being more malleable, wrought iron has high tensile strength and has been used for lattice girders, catenaries and cables, which all need to withstand the pressure of stretching.

The invention of mass-produced, high-quality steel made the engineer's job much easier, as steel combines the compressive strength of cast iron with the tensile strength of wrought iron. Steel therefore replaced iron in almost all bridges constructed from the end of the nineteenth century onwards. Concrete on its own, like stone, has poor tensile strength and is only used for abutments, piers and foundations. It is combined with steel to form reinforced concrete and pre-stressed concrete, and this allows much wider and flatter spans to be constructed than would be possible with ordinary concrete or with stone.

Notes

Chapter 1 Richmond and Twickenham

1. Held in Richmond Libraries Local Studies Collection
2. Richard Crisp, *Richmond and its Inhabitants from Olden Times* (Sampson, Low and Marston, 1866)
3. James Boswell, *Life of Johnson* (Richard Clay & Son, 1894)

Chapter 2 Kew

4. E.B. Chancellor, *An Account of the Bridges across the Thames at Kew* (J.H. Broad, 1903)
5. Horace Walpole, *Journal of the reign of King George the Third, from the year 1771 to 1783* (London, 1859)
6. Charlotte Papendiek, *Court and private life in the time of Queen Charlotte* (Bentley & Son, 1887)
7. E.B. Chancellor, *An Account of the Bridges across the Thames at Kew, op. cit.*

Chapter 3 Chiswick and Barnes

8. A.W. Winch, *Bits about Barnes* (London, 1895)

Chapter 4 Hammersmith

9. Thomas Faulkner, *The History and Antiquities of the Parish of Hammersmith* (T. Faulkner, 1839)

10. Daniel Defoe, *A Tour through the whole Island of Great Britain* (D. Browne, 1762)

Chapter 5 Putney and Wandsworth
11. Thomas Faulkner, *The History and Antiquities of the Parish of Hammersmith, op. cit.*
12. Archibald Chasmore, *The Old Bridge* (T.C. Davidson, 1875)
13. Sir Joseph W. Bazalgette, *An account of the metropolitan bridges over the Thames* (Metropolitan Board of Works, 1880)
14. George Dewe, *Fulham Bridge 1729–1886* (Fulham and Hammersmith Historical Society, 1986)

Chapter 6 Battersea and Chelsea
15. Minutes of Battersea Bridge Company 1776–93 (held in Wandsworth Borough Local History Library)
16. George Bryan, *Chelsea in the Olden and Present Times* (George Bryan, 1869)
17. Ceremonial pamphlet on the opening of Chelsea Bridge (LCC, 1937)
18. Tobias Smollett, *The expedition of Humphry Clinker* (J. Walker & Co., 1815)
19. Quoted in Aston Webb (ed.), *London of the Future* (T. Fisher Unwin, 1921)

Chapter 7 Vauxhall and Lambeth
20. Charles Hollis, *Proposed Improvements in Lambeth and Westminster* (William Clowes, 1829)
21. Thomas Allen, *The History and Antiquities of the Parish of Lambeth* (J. Allen, 1827)

Chapter 8 Westminster
22. Quoted in M.E.C. Walcott, *Westminster* (J. Masters, 1849)
23. *Journal of the Common Council* (5 October 1664, held in Guidhall Library)
24. Charles Labelye, *A Description of Westminster Bridge* (W. Strahan, 1751)
25. *Ibid.*

Chapter 9 Charing Cross
26. George Gater, *The Survey of London* (LCC, 1937)

Chapter 10 Waterloo
27. Quoted in E.B. Chancellor, *The Annals of the Strand* (Chapman & Hall, 1912)
28. Thomas Pennant, *Some Account of London* (R. Faulder, 1791)
29. Quoted in Aston Webb (ed.), *London of the Future, op. cit.*
30. Samuel Smiles, *Lives of the Engineers* (John Murray, 1874)
31. John Timbs, *The Romance of London* (Frederick Warne & Co., 1928)
32. Mark Searle, *Turnpikes and Toll-Bars* (Hutchinson & Co., 1930)
33. *The Report of the Royal Commission on Cross-river Traffic in London* (HMSO, 1926)
34. Quoted in Wallace Rayburn, *Bridge across the Atlantic* (Harrap, 1972)

Chapter 11 Blackfriars
35. John Stow, *A Survey of London* (Clarendon Press, 1908)
36. 'A Description of a City Shower', quoted in John Ashton, *The fleet, its river, prison, and marriages* (T. Fisher Unwin, 1888)
37. Publicus, *Observations on Bridge Building* (J. Townsend, 1760)
38. James Boswell, *Life of Johnson, op. cit.*
39. Anon., *The City Inscription on Pitt-Bridge, Blackfriars* (London, 1760)
40. J. Paterson, *A Plan to Raise £300,000* (London, 1767)
41. Quoted in A.E. Richardson, *Robert Mylne, Architect and Engineer* (B.T. Batsford, 1955)
42. W. Thornbury and E. Walford, *Old and New London* (Cassell & Co., 1897)
43. William Lucey, *The cost of new Blackfriars Bridge* (Waterlow & Sons, 1862)
44. Adrian Gray, *The London, Chatham & Dover Railway* (Meresborough Books, 1984)

Chapter 12 Millennium Bridge
45. Deyan Sudjic, *Blade of Light* (Penguin Press, 2001)

Chapter 13 Southwark and Cannon Street

46. John Rennie, *Autobiography* (E.F.N. Spon, 1875)
47. A Subscriber, *Considerations on the proposed Southwark Bridge* (Sherwood, Neely & Jones, 1813)
48. John Stow, *A Survey of London, op. cit.*
49. Alan Jackson, *London's Termini* (David & Charles, 1985)

Chapter 14 London Bridge

50. Snorri Sturluson, *Heimskringla: The Olaf Sagas*, trans. Samuel Laing (J.M. Dent, 1915)
51. John Stow, *Survey of London, op. cit.*
52. Gordon Home, *Old London Bridge* (Lane, 1931)
53. Patricia Pierce, *Old London Bridge* (Headline, 2001)
54. Thomas Pennant, *Some Account of London, op. cit.*
55. H. Chamberlain, *A New and compleat History and Survey of the Cities of London and Westminster* (J. Cooke, 1770)
56. C.W. Shepherd, *A Thousand Years of London Bridge* (Baker, 1971)

Chapter 15 Tower Bridge

57. W. Senior *et al.*, *The Royal River* (Cassell & Co., 1885)
58. John Wolfe Barry, *The Tower Bridge* (Boot, Son and Carpenter, 1894)
59. J.E. Tuit, *The Tower Bridge: its history and construction* (The Engineer, 1894)
60. Honor Godfrey, *Tower Bridge* (John Murray, 1988)

Sources

The research upon which this book is based has mainly been done in London's local and national libraries. Of national libraries, I have made extensive use of the resources of the British Library, the Newspaper Library at Collingdale and the House of Lords Record Office. I have also spent much time delving into publications and local material in the local archive libraries of the Corporation of London and the London boroughs that border the River Thames. The specific libraries I have used are recorded in the acknowledgements.

General sources
The following sources have provided much of the information on subjects relevant to many or all of the bridges.

Newspapers, journals and magazines
The Newspaper Library holds daily and weekly publications, but if you are looking for a journal, I would recommend checking whether it is held there or at the British Library before making a visit. Many local libraries also hold old newspapers and journals. Most of the articles cover a specific bridge, as referenced in the text of the book. I include a list here of publications which I have found useful on several occasions:

The Builder
Country Life
Daily Telegraph
The Engineer
Evening Standard
The Gentleman's Magazine
Illustrated London News
The Institution of Civil Engineers [ICE] Minutes of Proceedings
The Times

Books

I have used the following publications for background information on the history of London and the River Thames, on bridges and bridge building, on the railways of south London and for biographies of engineers and architects:

Bazalgette, Sir Joseph W., *An account of the metropolitan bridges over the Thames* (Metropolitan Board of Works, 1880)

Boswell, James, *Life of Johnson* (Richard Clay & Son, 1894)

Course, Edwin, *London's Railways: Then and Now* (Batsford, 1987)

Defoe, Daniel, *A Tour through the whole Island of Great Britain* (D. Browne, 1762)

De Mare, Eric, *Bridges of Britain* (Batsford, 1975)

Dredge, J., *Thames Bridges from the Tower to the Source* (reproduced from *Engineering*, 1896–8)

Herring, J.H., *Thames Bridges from London to Hampton Court* (H.R. Pinder, 1884)

Hopkins, H.J., *A Span of Bridges* (David & Charles, 1970)

Jackson, Alan, *London's Termini* (David & Charles, 1985)

Maitland, W., *History and Survey of London* (London, 1756)

Marshall, C.F.D., *A History of the Southern Railway* (Ian Allen, 1963)

Nicholls, C.S. (ed.), *Dictionary of National Biography* (Oxford University Press, 1996)

Pannell, J.P.M., *An Illustrated History of Civil Engineering* (Thames & Hudson, 1964)

Philips, Geoffrey, *Thames Crossings* (David & Charles, 1981)

Phillips, Hugh, *The Thames about 1750* (Collins, 1951)

Pudney, John, *Crossing London's River* (Dent, 1972)

Rolt, L.T.C., *Victorian Engineering* (Penguin, 1988)

Ruddock, Ted, *Arch Bridges and their Builders* (Cambridge University Press, 1979)

Searle, Mark, *Turnpikes and Toll-Bars* (Hutchinson & Co., 1930)

Senior, W. *et al.*, *The Royal River* (Cassell & Co., 1885)

Smiles, Samuel, *Lives of the Engineers* (John Murray, 1874)

Stow, John, *A Survey of London* (Clarendon Press, 1908)

Survey of London, The (London Survey Committee, 1900)

Timbs, John, *Curiosities of London* (John Camden Hotten, 1871)

Thornbury, W. and E. Walford, *Old and New London* (Cassell & Co., 1897)

Weinreb, Ben and Christopher Hibbert, *The London Encyclopaedia* (Macmillan, 1995)

White, H.P., *The Regional History of Railways of Great Britain*, Vol. 3 (David & Charles, 1971)

Official documents

Much detailed information has been gleaned from Acts of Parliament, select committee reports and Metropolitan Board of Works reports. The most significant Acts of Parliament for individual bridges are referenced under the relevant chapter. The following items cover several bridges and have been referred to a number of times in the text:

The Report of the Metropolitan Commission on Thames Bridges (HMSO, 1857)

The Metropolis Toll Bridges Act 40 & 41 Vic. I c. CXCII 1877

The Report of the Royal Commission on Cross-river Traffic in London (HMSO, 1926)

Sources for individual chapters

The main sources are listed here, including the most important Acts of Parliament. Press cuttings and articles from local and national newspapers and journals have provided additional information and anecdotes.

Chapter 1
Acts of Parliament
Richmond Bridge Act 13 Geo. III c. 83 1773

Publications

Cloake, John, *Richmond Past: A Visual History of Richmond, Kew, Petersham and Ham* (Historical Publications, 1991)

Crisp, Richard, *Richmond and its Inhabitants from Olden Times* (Sampson, Low and Marston, 1866)

Dunbar, Janet, *A Prospect of Richmond* (White Lion Publishers, 1973)

Howard, Diana, *Richmond Bridge and other Thames crossings between Hampton and Barnes* (London Borough of Richmond, 1976)

Matthews, Simon, *Richmond Bridge Bicentenary 1777–1977* (C. James, 1980)

Ransomes and Rapier, *Richmond and Twickenham footbridge, lock and weir* (Waterlow & Sons Ltd, 1894)

Richmond Libraries Local Studies Collection archive material

Ceremonial programme for the opening of the Thames Bridges between Middlesex and Surrey (3 July 1933)

Ceremonial programme for the centenary of Richmond and Twickenham Footbridge, Lock and Weir (PLA, 1994)

Dalrymple-Hay, H.H., article in *The Engineer* on the widening of Richmond Bridge (16 April 1937)

Illustrated London News articles on Richmond Railway Bridge (21 October 1848 and 15 June 1867)

Maxwell, Donald, undated untitled article

Richmond and Twickenham Times articles

Richmond Bridge commissioners' minute book 1773–86 (manuscript)

Chapter 2
Acts of Parliament
Kew Bridge Act 30 Geo. II c. 63 1757
Kew Bridge Act 22 Geo. III c. 42 1782
Kew Bridge Act 61 & 62 Vic. I c. CLV 1898

Barnard, John, *Design of a bridge over the Thames at Kew* (London, 1759)

Blomfield, David, *Kew Past* (Phillimore & Co., 1994)

Chancellor, E.B., *An Account of the Bridges across the Thames at Kew* (J.H. Broad, 1903)

Dickens, Charles, *Dictionary of the Thames* (Charles Dickens, 1879)

Jerome, Jerome K., *Three Men in a Boat* (Readers Library Publishing Co., 1929)

Thacker, E.S., *Thames Highway* (David & Charles, 1968)

Turner, Fred, *The history and antiquities of Brentford* (Walter Pearce & Co., 1922)

Richmond Libraries Local Studies Collection, and Chiswick and Hounslow Libraries Local Studies Collection archive material
The Engineer article on Kew Railway Bridge (16 April 1869)
Kew Bridge centenary display catalogue (Richmond Library, 2003)
Richmond History Journal article on Kew Bridge (Issue 25, 2004)

Chapter 3
Acts of Parliament
Middlesex and Surrey (Thames Bridges &c.) Act 13 Geo. V c. LXXII 1928

Hammersmith and Fulham Archives and Local History Centre archive material
Barnes and Mortlake History Society Newsletter
Ceremonial programme for the opening of the Thames bridges between Middlesex and Surrey (3 July 1933)
Crutwell, G.E., 'The new bridge of the LCDR Company at Blackfriars', in *ICE Minutes of the Proceedings* Vol. 101 (1889–90)
Roads and Road Construction Vol. 10 (June 1932)
Szlumper, A.W., 'Barnes Bridge', in *ICE Minutes of the Proceedings* Vol. 124 (1895–6)

Chapter 4
Acts of Parliament
Hammersmith Bridge Act 5 Geo. IV c. CXII 1824
Hammersmith Bridge Act 9 Geo. IV c. LII 1828
Hammersmith Bridge Act 46 & 47 Vic. I c. CLXXVII 1883

Publications
The history of Hammersmith Bridge up to 1987 has been covered in detail in Charles Hailstone's comprehensively researched *Hammersmith Bridge* (E.H. Baker & Co. Ltd, 1987), and I am indebted to him for much of my material.

I have also consulted the following:
Faulkner, Thomas, *The History and Antiquities of the Parish of Hammersmith* (T. Faulkner, 1839)
Opening of Hammersmith Bridge (MBW, 1887)
Whiting, Philip D., *A history of Hammersmith based upon that of Thomas Faulkner in 1839* (Hammersmith Local History Group, 1965)

Hammersmith and Fulham Archives and Local History Centre archive material
Cumming, T.G., *Description of the iron bridges of suspension over the Straits of Menai and over the River Thames at Hammersmith* (J. Taylor, 1828)
Franklin Journal and American Mechanics' Magazine article on Hammersmith Bridge (April 1828)
Hammersmith Bridge Company records
West London Observer articles

Chapter 5
Acts of Parliament
Fulham Bridge Act 12 Geo. I c. 36 1725–6
Fulham Bridge Act 1 Geo. II c. 18 1727–8
Metropolitan Bridges Act 44 & 45 Vic. I c. CXCII 1881
Wandsworth Bridge Act 27 & 28 Vic. I c. CCXXXVIII 1884

Publications

Bowack, John, *The antiquities of Middlesex*, Vol. 2 (S. Keble, 1706)

Chasmore, Archibald, *The Old Bridge* (T.C. Davidson, 1875)

Croker, Thomas Crofton, *A Walk from London to Fulham* (William Tegg, 1860)

Dewe, George, *Fulham Bridge 1729–1886* (Fulham and Hammersmith Historical Society, 1986)

Faulkner, Thomas, *The History and Antiquities of the Parish of Hammersmith* (T. Faulkner, 1813)

Feret, Charles, *Fulham old and new* (Leadenhall Press, 1900)

Gerhold, Dorian, *Putney and Roehampton Past* (Historical Publications, 1994)

Wandsworth Local History Service archive material

Jones, T.E., 'A short history of the three bridges over the Thames within the Fulham district' (1880)

Wandsworth History Society News Sheet 1 (1969)

Articles and press cuttings from:

History Today

Putney and Wandsworth Borough News

Wandsworth Borough News

Wandsworth Historian

Chapter 6

Acts of Parliament

Battersea Bridge Act 6 Geo. 3 c. 66 1766

Chelsea Bridge Act 9 & 10 Vic. I c. 112 1846

Metropolitan Bridges Act 44 & 45 Vic. I c. CXCII 1881

Albert Bridge Act 27 & 28 Vic. I c. CCXXXV 1864

Metropolitan Bridges Act 47 & 48 Vic. I c. CCXXVIII 1884

Publications

Bryan, George, *Chelsea in the Olden and Present Times* (George Bryan, 1869)

Buckton, E.J., 'Chelsea Bridge reconstruction', in *ICE Journal* Vol. 7 No. 3 (January 1938)

Denny, Barbara, *Chelsea Past* (Historical Publications, 1996)

Simmonds, H.S., *All about Battersea* (Ashfield, 1882)

Taylor, J.G., *Our Lady of Battersea* (George White, 1925)

Walker, Annabel, *Kensington and Chelsea* (John Murray, 1987)

Kensington and Chelsea Local Studies Collection archive material

Ceremonial pamphlet on the opening of Chelsea Bridge (LCC, 1937)

Illustrated London News article on Albert Bridge (23 August 1873)

Illustrated London News article on Chelsea Bridge (25 September 1858)

London Metropolitan Archives archive material

Proof of evidence, Plans and Proceedings regarding the Public Inquiry into the proposed closure of Albert Bridge (HMSO, 1974)

Wandsworth Local History Service archive material

Minutes of Battersea Bridge Company 1776–93

'The opening, free of toll, of Wandsworth, Putney (otherwise Fulham) and Hammersmith Bridges' (MBW, June 1880)

South London Press, Putney and Wandsworth Borough News and *Wandsworth Borough News* articles

Chapter 7

Acts of Parliament

Vauxhall Bridge Act 49 Geo. III c. CXLII 1809

Lambeth Bridge Act 24 & 25 Vic. I c. CXVII 1861

Vauxhall Bridge Act 58 & 59 Vic. I c. CXXIX 1895

Publications

Allen, Thomas, *The History and Antiquities of the Parish of Lambeth* (J. Allen, 1827)

Copperthwaite, W.C., 'Vauxhall Bridge', in *ICE Minutes of the Proceedings* Vol. 169 (1906–7)

Dodd, Ralph, *Introductory Report on the intended Bridge across the River Thames, from near Vauxhall to the opposite shore, etc.* (London, 1810)

Fox, C.D., 'The widening of Victoria Bridge', in *ICE Minutes of the Proceedings*, Vol. 27 (1867–8)

Groves, G.L., 'Lambeth Bridge', in *ICE Minutes of the Proceedings*, Vol. 239 (1934–5)

Hollis, Charles, *Proposed Improvements in Lambeth and Westminster* (William Clowes, 1829)

Kumar, A., 'Lambeth Bridge strengthening', in *ICE Minutes of the Proceedings*, Vol. 156 (May 2003)

Robbins, Michael and T.C. Barker, *The History of London Transport*, Vol. 1 (Allen & Unwin, 1963)

Watson, Elizabeth, *Westminster and Pimlico Past* (Historical Publications, 1993)

Lambeth Archives material

Programme for the opening of Lambeth Bridge (LCC, 1932)

'Programme and plan of route for the procession on the occasion of the opening, free of toll, of Lambeth, Vauxhall, Chelsea, Albert and Battersea Bridges' (MBW, 1879)

Report of the Select Committee on Lambeth Bridge (House of Commons, 1928)

Chapter 8

Acts of Parliament

Westminster Bridge Act 9 Geo. II c. 29 1735–6

Westminster Bridge Act 16 & 17 Vic. I c. 46 1853

Publications

Hunting, Penelope, *Royal Westminster* (RICS, 1981)

Labelye, Charles, *A Description of Westminster Bridge* (W. Strahan, 1751)

Langley, Batty, *A Survey of Westminster Bridge as 'tis now sinking into ruin* (M. Cooper, 1748)

Walcott, M.E.C., *Westminster* (J. Masters, 1849)

Walker, R.J.B., *Old Westminster Bridge* (David & Charles, 1979)

Watson, Elizabeth, *Westminster and Pimlico Past* (Historical Publications, 1993)

City of Westminster Archives Centre material
Carson, P., *Journal of Transport History* article on Westminster
    Bridge (1957)
Reports of the select committees on Westminster Bridge
    (1844–57)

Chapter 9
Acts of Parliament
Hungerford Bridge Act 6 & 7 William IV c. CXXXIII 1836
Hungerford Railway Bridge Act 27 & 28 Vic. I c. LXXXI 1859

Publications
Anon., *Descriptive particulars of Hungerford suspension bridge*
    (London, 1845)
Gater, George, *The Survey of London* (LCC, 1937)
Keen, Arthur, *Charing Cross Bridge* (Ernest Benn, 1930)
MOT London and Home Counties Traffic Advisory Committee,
    Report on Charing Cross Bridge (HMSO, 1936)
Parker, J. *et al.*, 'Hungerford Bridge Millennium project', in *ICE
    Minutes of the Proceedings*, Issue CE (May 2003)
Timbs, John, *The Romance of London* (Frederick Warne & Co., 1928)

City of Westminster Archives Centre material
Review of the Charing Cross Bridge Scheme (City of Westminster,
    1935)

Chapter 10
Acts of Parliament
Strand Bridge Act 49 Geo. III c. CXCI 1809
Strand Bridge Act 56 Geo. III c. LXIII 1816

Publications
Bruce, J., *The History of the Ancient Savoy Palace, built by the Duke de
    Savoy, A.D. 1245, now the site of the Waterloo Bridge* (Sherwood,
    Neely & Jones, 1817)
Buckton, E.J., 'The demolition of Waterloo Bridge', in *ICE
    Journal*, Vol. 3 No. 8 (1936)

Buckton, E.J., 'The new Waterloo Bridge', in *ICE Journal*, Vol. 20 No. 7 (1943)

Elliot, M., *Waterloo Bridge: Its Swansong* (Gibbs, Bemwold & Co., 1934)

Rayburn, Wallace, *Bridge across the Atlantic* (Harrap, 1972)

Waterloo Bridge Opening (LCC, 1945)

City of Westminster Archives Centre material
Pamphlets on the LCC Strand underpass and the archaeology of the Strand

Chapter 11
Acts of Parliament
Blackfriars Bridge Act 29 Geo. II c. 86 1755–6
Blackfriars Bridge Act 26 & 27 Vic. I c. LXII 1863
City of London (Blackfriars and other Bridges) Act 5 Edw. VII c. CLXXX 1906

Publications
Anderson, D. and H. Cunningham, 'The widening of Blackfriars Bridge', in *ICE Minutes of the Proceedings*, Vol. 190 (1911–12)

Ashton, J., *The fleet, its river, prison, and marriages* (T. Fisher Unwin, 1888)

Crutwell, G.E., 'The new bridge of the LCDR Company at Blackfriars', in *ICE Minutes of the Proceedings*, Vol. 101(1889–90)

Glossop, Rudolph, *Blackfriars Bridge under Construction* (Mowlem, 1965)

Gray, Adrian, *The London, Chatham & Dover Railway* (Meresborough Books, 1984)

Lucey, William, *The cost of new Blackfriars Bridge* (Waterlow & Sons, 1862)

Paterson, J., *A Plan to Raise £300,000* (London, 1767)

Publicus, *Observations on Bridge Building* (J. Townsend, 1760)

Richardson, A.E., *Robert Mylne, Architect and Engineer* (B.T. Batsford, 1955)

Guildhall and London Metropolitan Archives material
Blackfriars Bridge Committee minute book 1772–8 and 1778–84
City of London Common Council Blackfriars Bridge Reports
(1754, 1759, 1771, 1776, 1784, 1786, 1840, 1841, 1909, 1911)

Chapter 12
Publications
Sudjic, Deyan, *Blade of Light* (Penguin Press, 2001)

Southwark Local History Library archive material
Miscellaneous press articles from local and national newspapers

Chapter 13
Acts of Parliament
Southwark Bridge Act 51 Geo. III c. 166 1811
Corporation of London (Bridges) Act 2 Geo. V c. CXX 1911

Publications
Jackson, Alan, *London's Termini* (David & Charles, 1985)
Rennie, John, *Autobiography* (E.F.N. Spon, 1875)

Guildhall Library and Southwark Local History Library archive
    material
Anon., *Some Account of the Southwark Bridge* (Nicholas, Son and
    Bentley, 1917)
Bridge House Estates Committee reports (City of London, 1903,
    1910)
Lippard, D.C., 'Cannon Street Station air rights development',
    in *ICE Minutes of the Proceedings*, Vol. 99 (1993)
Southwark Bridge (City of London, 1921)
Subscriber, A, *Considerations on the proposed Southwark Bridge*
    (Sherwood, Neely & Jones, 1813)

Chapter 14
Acts of Parliament
London Bridge Act 29 Geo. II c. 40 1756
London Bridge Act 4 Geo. IV c. 50 1823
London Bridge Act 1967

Publications
Chamberlain, H., *A New and compleat History and Survey of the Cities of London and Westminster* (J. Cooke, 1770)
City of London Engineers Department, *The Story of Three Bridges: The History of London Bridge* (Finsbury Publishing Ltd, 1973)
Home, Gordon, *Old London Bridge* (Lane, 1931)
Jackson, Peter, *London Bridge* (Cassell, 1971)
Pennant, Thomas, *Some Account of London* (R. Faulder, 1791)
Pierce, Patricia, *Old London Bridge* (Headline, 2001)
Shepherd, C.W., *A Thousand Years of London Bridge* (Baker, 1971)

Chapter 15
Acts of Parliament
Corporation of London (Tower Bridge) Act 48 & 49 Vic. I c. CXCV 1885

Publications
Barry, John Wolfe, *The Tower Bridge* (Boot, Son and Carpenter, 1894)
Crutwell, G.E.W., *The Tower Bridge: Superstructure* (ICE, 1897)
Godfrey, Honor, *Tower Bridge* (John Murray, 1988)
Tuit, J.E., *The Tower Bridge: its history and construction* (The Engineer, 1894)
Vallintine, Reg, *Divers and Diving* (Blandford, 1981)
Welch, Charles, *The History of Tower Bridge and other Bridges over the Thames* (Smith, Elder, 1894)

Tower Bridge Exhibition
This permanent exhibition housed in the bridge itself is especially useful for information on the walkways and the operation of the bascules.

Index